COFFEE

Volume 2: Technology

COFFEE

Volume 1: Chemistry
Volume 2: Technology
Volume 3: Physiology
Volume 4: Agronomy
Volume 5: Related Beverages
Volume 6: Commercial and Technico-Legal Aspects

COFFEE

Volume 2: TECHNOLOGY

Edited by

R. J. CLARKE

Formerly of General Foods Ltd, Banbury, UK

and

R. MACRAE

Department of Food Science, University of Reading, UK

ELSEVIER APPLIED SCIENCE
LONDON and NEW YORK

ELSEVIER SCIENCE PUBLISHERS LTD
Crown House, Linton Road, Barking, Essex IG11 8JU, England

Sole Distributor in the USA and Canada
ELSEVIER SCIENCE PUBLISHING CO., INC.
655 Avenue of the Americas, New York, NY 10010, USA

WITH 42 TABLES AND 81 ILLUSTRATIONS

© 1987 ELSEVIER SCIENCE PUBLISHERS LTD
First Edition 1987
Reprinted 1989

British Library Cataloguing in Publication Data

Coffee.
 Vol. 2: Technology
 1. Coffee
 I. Clarke, R. J.   II. Macrae, R.
 641.3'373   TX415

ISBN 978-94-010-8028-6   ISBN 978-94-009-3417-7 (eBook)
DOI 10.1007/978-94-009-3417-7

Library of Congress CIP data applied for

No responsibility is assumed by the Publisher for any injury and/or damage to persons or property as a matter of products liability, negligence or otherwise, or from any use or operation of any methods, products, instructions or ideas contained in the material herein.

Special regulations for readers in the USA

This publication has been registered with the Copyright Clearance Center Inc. (CCC), Salem, Massachusetts. Information can be obtained from the CCC about conditions under which photocopies of parts of this publication may be made in the USA. All other copyright questions, including photocopying outside of the USA, should be referred to the publisher.

All rights reserved. No part of this publication may be reproduced, stored in a retrieval system, or transmitted in any form or by any means, electronic, mechanical, photocopying, recording, or otherwise, without the prior written permission of the publisher.

Preface

The present volume, Volume 2 in this planned series on coffee, deals with processing and follows on naturally from the first volume on the chemistry of coffee, which described its numerous constituents in the green (raw) and various product forms.

We have already remarked that coffee has great compositional complexity, and this complexity of understanding extends when we come to consider its processing; that is, the many processes involved in the roasting of green coffee and its subsequent conversion into a consumable brew, especially through extraction and drying into an instant coffee. The simple brewing of roasted and ground coffee with water in the home also possesses considerable mystique and needs know-how for optimal results. The choice of green coffees from an almost bewildering array of different types available, through species/variety differences and different methods of processing from the coffee cherry to the green coffee bean, needs understanding and guidance. Furthermore, various forms of pre-treatment of green coffee before roasting are available. Some of these are little known, but others such as decaffeination, for those who desire roasted or instant coffee with little or no caffeine, are now becoming well established. Finally, both the processing of coffee cherries to coffee beans, leaving a range of different waste products (pulp, hulls, husk, parchment, etc.), and of roasted coffee after industrial aqueous extraction, leaving spent coffee grounds, provide waste products that have found considerable commercial value in different ways.

In our nine chapters, therefore, all these subjects are dealt with in detail, bringing together much information that has not generally been available previously in the English language. Particularly, in the basic unit operations of roasting–grinding–extraction–drying, the approach has been that of considering the fundamentals of the subject, through the well established discipline of chemical (food) engineering. This approach to coffee processing has been applied by a number of workers in the last two decades as described in the Appendix (section 5); though, as will be seen, much remains to be revealed and understood in this complex field. Much information and opinion is empirical, so that, particularly in instant coffee manufacture, there are many elements of craft, skill and know-how in plant operation. However, it is not the aim of this volume to cover these facets, and indeed much of this is necessarily known only to the manufacturers themselves; though an indication of preferred or 'wished-for' modes of operation can be seen on studying the numerous patents (probably some separate 1000) on coffee that have been granted. Only a selection of these patents can be given in this volume, with a preference towards noting the patents granted in the USA, though many of these may well also be issued in other countries. There is no doubt that the intense competition among the major instant coffee manufacturers of the world has been the spur to great progress in the quality of coffee products over the last three or four decades.

A vexing problem common now to all *technical* texts on the engineering of food and other manufacturing processes is in the choice of units for the various physical properties of substances and process parameters involved. Much existing information is expressed in so-called engineering units (foot-pound-hour system), especially in the USA; and movement to full use of the metric system (centimetre-gram-second system) has been generally slow in English-speaking countries, except in reporting results of laboratory experimentation. Furthermore, we are now being encouraged to move further, to the adoption of the SI system (Système Internationale) of units (metre-kilogram-second), which is the most logical of all. In this volume, the metric system is used as the basis, including for operations on the industrial scale, though simultaneous conversion is made back to engineering units where information arose in this system originally. Reference is also made to the corresponding SI units where it is thought to be appropriate, and the important relationships between the various relevant units used at different times are set out in the Appendix. A similar problem arises in the choice of symbols for physical properties and parameters; though a common system is developing, in discussion of the work of certain authors, this volume also uses the particular symbols as used originally.

Each chapter has been written by an international expert in that particular field. It is therefore to be hoped that the present volume will provide a convenient and readable source of information and reference in the English language for all those interested in the industrial processing of coffee, and in the remaining but important process left to the consumer, i.e. coffee brewing.

R. J. CLARKE
R. MACRAE

Contents

Preface v

List of Contributors xiv

Chapter 1 *Green Coffee Processing*   J-C. VINCENT
1. Introduction. 1
2. Dry Processing Method 3
   2.1. Natural drying 3
   2.2. Artificial drying 4
3. Wet Processing Method 8
   3.1. Receiving 10
   3.2. Pulping 10
   3.3. Separation/classification. 15
   3.4. Fermentation. 17
   3.5. Washing 18
   3.6. Draining and pre-drying 19
   3.7. Drying of the parchment coffee 19
4. Curing 22
   4.1. Redrying 22
   4.2. Cleaning 22
   4.3. Hulling 22
   4.4. Size grading 26
   4.5. Density sorting 27
   4.6. Colorimetric sorting 28
5. Storage 31
6. Handling 32
   References 33

Chapter 2 Grading, Storage, Pre-treatments and Blending
R. J. CLARKE

1. Introduction. 35
2. Marketed Grades. 35
   - 2.1. Systems of specification. 36
   - 2.2. Liquoring/flavour characteristics 38
   - 2.3. Bean size and shape 39
   - 2.4. Defects 42
   - 2.5. Colour 48
   - 2.6. Roasting characteristics. 48
   - 2.7. Bulk density 49
   - 2.8. Crop year 49
3. Storage 49
   - 3.1. Storage conditions. 49
   - 3.2. Isotherms 50
   - 3.3. Storage stability 50
   - 3.4. Methods of storage 51
4. Pre-treatments 52
   - 4.1. Cleaning and destoning 52
   - 4.2. 'Health' coffees 53
5. Selection and Blending. 54
   - 5.1. Availability 54
   - 5.2. Selection 54
   - 5.3. Blending methods 57
   - References 57

Chapter 3 Decaffeination of Coffee   S. N. KATZ

1. Introduction. 59
2. Solvent Decaffeination 61
3. Water Decaffeination 64
4. Supercritical Carbon Dioxide Decaffeination . . . 66
5. Decaffeination of Roasted Coffee and Extract . . . 68
6. Caffeine Refining 69
   - References 70

Chapter 4 Roasting and Grinding   R. J. CLARKE

1. Introduction. 73
2. Process Factors in Roasting. 73
   - 2.1. Mechanisms and methods 73
   - 2.2. Chemical changes 75
   - 2.3. Heat factors 79
   - 2.4. Physical changes 84
   - 2.5. Measurement of roast degree. 86
   - 2.6. Emission control of organic compounds and chaff. . 87
3. Roasting Equipment 89
   - 3.1. Horizontal drum roasters 89

3.2. Vertical fixed drum with paddles 93
3.3. Rotating bowl 95
3.4. Fluidised beds 96
3.5. Pressure operation 97
3.6. Roaster ancillaries 97
4. Process Factors in Grinding 97
  4.1. Mechanism of grinding 97
  4.2. Size analysis 99
  4.3. Bulk density of ground coffee 104
5. Grinding Equipment 104
References 106

Chapter 5   Extraction   R. J. CLARKE
1. Introduction. 109
2. Mechanisms and Methods 110
  2.1. Methods 110
  2.2. Mechanism of soluble solids extraction 114
  2.3. Mechanism of volatile compound extraction . . . 115
  2.4. Compositional factors 117
  2.5. Balances and rate and productivity factors . . . 122
  2.6. Volatile compound handling 134
3. Process Equipment 141
  3.1. Percolation batteries 141
  3.2. Continuous countercurrent screw extractor . . . 142
  3.3. Process control measurements 143
References 144

Chapter 6   Drying   R. J. CLARKE
1. Introduction. 147
2. Process Factors in Spray-drying 149
  2.1. Methods 149
  2.2. Compositional changes 150
  2.3. Spray formation 151
  2.4. Spray–air contact 155
  2.5. Mechanisms of water removal 156
  2.6. Mechanisms of volatile compound retention . . . 167
  2.7. Fines separation 179
  2.8. Agglomeration 180
3. Process Factors in Freeze-drying 181
  3.1. Methods 181
  3.2. Mechanism of water removal. 182
  3.3. Mechanism of volatile compound retention . . . 188
4. Process Factors in Pre-concentration 190
  4.1. Evaporation 190
  4.2. Freeze-concentration 190
  4.3. Reverse Osmosis 192

5. Process Equipment 192
   5.1. Spray driers 192
   5.2. Agglomerators 193
   5.3. Evaporators 194
   5.4. Freeze concentrators 194
   5.5. Freeze driers 194
   5.6. Other driers 195
   5.7. Process control 195
   5.8. Dust and fire hazards 196
   References 197

Chapter 7 *Packing of Roast and Instant Coffee* R. J. CLARKE
1. Introduction. 201
2. Packing of Roast Whole Bean Coffee 202
   2.1. Carbon dioxide evolution 202
   2.2. Stability factors 206
   2.3. Types of pack 206
3. Packing of Roast and Ground Coffee 207
   3.1. Carbon dioxide evolution 207
   3.2. Stability factors 209
   3.3. Types of pack 211
4. Packing of Instant Coffee 215
5. Packing Equipment 217
   5.1. Degassing plant 217
   5.2. Roast and ground coffee 218
   5.3. Instant coffee. 218
   5.4. Weight control 218
   References 219

Chapter 8 *Home and Catering Brewing of Coffee* G. PICTET
1. Introduction. 221
2. Bibliographic Review 222
   2.1. Solid–liquid extraction 222
   2.2. Brewing of roast coffee 223
   2.3. Properties of the coffee brew 230
3. Personal Research 234
   3.1. Introduction 234
   3.2. Experimental data 234
   3.3. Discussion of results 238
4. General Conclusions 250
   References 252

Chapter 9 *Waste Products* M. R. ADAMS and J. DOUGAN
1. Primary Processing: the Production of Green Coffee . . . 257
   1.1. Dry or natural processing 257
   1.2. Wet processing 259

2. Secondary Processing: the Production of Instant Coffee . . 283
   2.1. Coffee grounds 283
   2.2. Spent coffee grounds as fuel and feed 286
   References 287

Appendix
1. Units 293
   1.1. SI base units 293
   1.2. Some SI derived units used in engineering . . . 293
   1.3. Some prefixes for SI units 294
   1.4. Some conversions of SI and non-SI units . . . 295
   1.5. Dimensionless units used 296
2. Symbols for Physical Quantities in Equations . . . 298
3. Abbreviations 298
4. Flavour Terminology 299
5. Process Engineering Terminology. 300
   5.1. Food engineering and unit operations 300
6. Listing of British and International Standards Relating to Coffee 304

Index 307

List of Contributors

M. R. ADAMS

    Department of Microbiology, University of Surrey, Guildford, Surrey GU2 5XH, UK

R. J. CLARKE

    Ashby Cottage, Donnington, Chichester, Sussex PO20 7PW, UK

J. DOUGAN

    Tropical Development and Research Institute, 56–62 Gray's Inn Road, London WC1X 8LU, UK

S. N. KATZ

    General Foods Manufacturing Corporation, Maxwell House Division, 1125 Hudson Street, Hoboken, New Jersey 07030, USA

G. PICTET

    Linor—Food Development Centre, Nestec Ltd, Orbe, Switzerland

J-C. VINCENT

    Institut de Recherches du Café, du Cacao, Centre Gerdat, Avenue du Val de Montferrand, BP 5035, 34032 Montpellier Cedex, France

Chapter 1

Green Coffee Processing

J-C. VINCENT

Institut de Recherches du Café, du Cacao, Montpellier, France

1. INTRODUCTION

At the coffee production sites (farms and estates), two different main methods of processing are used to obtain intermediate products that will subsequently be treated in exactly the same way to provide the coffee beans of commerce. These methods are dry processing, which produces dried cherry coffee and wet processing, which produces (dry) parchment coffee. Dry processing is generally used for robusta coffee, but is also used in Brazil for the majority of arabica coffees. Wet processing, on the other hand, is used for arabica and results in so-called mild coffee, when fermentation is included in the preparation process. Dry processing is very simple and, most important of all, is less demanding in respect of harvesting, since all the berries or cherries are dried immediately after harvest. In contrast, wet processing requires more strict control of the harvesting as unripe berries or berries that have partly dried on the tree cannot be handled by the pulping machines.

After drying, the processing stages (or so-called curing) of dried cherry coffee and parchment coffee are very similar, differences lying in details of the equipment used to remove the remaining outer coverings from the green coffee beans, size grading and colorimetric sorting. The various stages of the overall processes are illustrated in Fig. 1. The preparation technique may not play a real role in the intrinsic quality of the final green coffee, but it is nevertheless significant if damage to the product is to be avoided. At present not enough attention is paid to the way in

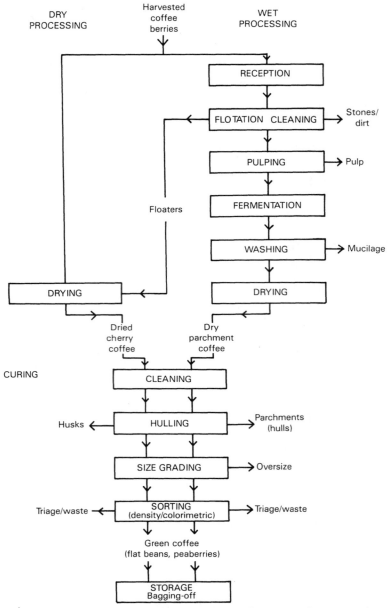

Fig. 1. Flow sheet illustrating the stages of wet and dry processing.

which the coffee is prepared, and it seems absurd to neglect the final preparation when so much care has gone into producing it. The other key element related to quality is the harvest: as is true of all fruits, only ripe fruits or berries have developed the optimal level of sugars and aroma precursors that will eventually be transformed during the roasting of the final green coffee beans.

2. DRY PROCESSING METHOD

Dry processing is the simplest method. It consists of drying the whole fruits immediately after harvesting. This method is used for robusta coffee, but also for 90% of Brazilian arabica coffees. It is highly convenient where there is not enough available labour to ensure that only ripe fruits are harvested. In order to reduce the number of 'runs' among the trees, there is a trend to pick the fruits by the 'strip' method. This means that ripe, unripe and overripe fruits are picked simultaneously; in certain instances, it is not unusual to find up to 80% of unripe fruits. Coffee picked in this way cannot produce a 'fine' brew (that is, after roasting and infusion with hot water); in fact, such coffee is often described as 'hard', though this term has a number of different connotations. Drying is carried out either 'naturally' or artificially, as described below.

2.1. Natural Drying

This means drying in the sun, which has the advantage of not requiring any investment in equipment. Large drying areas are necessary, on account of the quantity of material to be dried and the slowness of the drying process of the berries, the pulp of which contains sugars and pectins. Sun-drying is usually carried out on clean dry ground, trays or a solid concrete surface. Irrespective of the type of surface, the berries must be spread out in a thin layer only 30–40 mm thick, especially at the start of the drying period, so that fermentation will not occur; in fact, any undue heating at this stage can cause discoloration (brown silverskin) and an unsatisfactory brew. Frequent raking is necessary during the earlier wet stage to stop moulds proliferating. Drying areas should be located in well ventilated places, thus clearings in equatorial forests should not be used as there is not enough ventilation and as stagnant fogs can form.

Many micro-organisms can proliferate on the skin (exocarp), particularly moulds, either directly in the growing area (e.g. *Colletotrichum coffeanum, Fusarium*) or on drying surfaces (e.g. *Aspergillus niger, Penicillium* sp., *Rhizopus*, etc.). Certain yeasts (*Torula, Saccharomyces*, etc.) and certain bacteria can also develop.

For a given thickness of layer, the length of the drying process depends mainly on the weather conditions including the psychrometric characteristics of the ambient air. It also depends upon the size of the berries, their degree of ripeness and their moisture content. In Brazil, some estates first sort the berries under water according to their diameter in order to provide more homogeneous batches of berries for drying. Degree of ripeness is associated with their moisture content, and their capacity to float or not in water. Thus floating berries (overripe, dried) have 20–50% moisture, and non-floating (green, yellow and red) will have 50–70%. According to Medcalf,[1] the distribution is as indicated in Table 1.

The berries to be dried will also be of varying size since the volume of the fruit increases in the ripening process and subsequently diminishes during drying on the tree. The same author obtained the results in Table 2 from size grading studies.

The required length of the drying period varies from 3 to 4 weeks. In drying robusta berries, the bulk density of the berries ranging over 645 kg m^{-3} to 440 kg m^{-3}, with a layer 30 mm thick, the load will range from 19·4 to 13·2 kg m^{-2}. The moisture content of the berries is reduced after drying to approximately 12% (as is). At this value, it is generally considered possible to conserve the beans satisfactorily in subsequent storage.

2.2. Artificial Drying

Artificial drying is practised on large estates for the reasons listed previously, and also to limit the cost of labour. As is also true of many

Table 1
Typical distribution of non-floating and floating coffee berries at different degrees of ripeness/moisture content

Category	Non-floating	Floating
Green	100	0
Unripe	94	6
Ripe	97	3
Overripe	88	12
Dried cherries	36	64
Husks	0	100

Table 2
Typical size distribution of coffee berries at different degrees of ripeness

Size of berries (small diameter in mm)	Percentage by weight		
	Ripe	Overripe	Dried
12	51	31	3
11	31	30	11
10	16	35	56
9·0	2	4	23
7·8	0	0	7

food products, the drying temperature is the limiting factor in the allowable speed of drying. More particularly, it is known that very high temperatures at the start of the drying period can lead to the formation of so-called 'stinker' beans, a very undesirable defect in green coffee. Temperature conditions need to be even more strictly controlled in drying cherry coffee than parchment (see page 19).

Various authors recommend drying at low temperatures, e.g. 55°C according to Roelofsen.[2] Our own tests confirm that a mediocre product is obtained (i.e. reddish beans with brown silverskin, responsible for unpleasant to stinking brews) if high temperatures are used; the same is true even at 60°C if there is insufficient airflow. Even in the case of dried coffee cherry at low moisture content, it is better to avoid high temperatures; above 80°C (temperature of ingoing air), there is a deterioration of taste. The use of low temperatures is, however, restrictive in that there is a noticeable increase in the length of the drying period below 60°C, and the thermodynamic position may become unfavourable. It is important to distinguish between actual coffee temperature, which can be measured though with some difficulty, and air temperatures (inlet and outlet), which are readily measured. As the former is of more significance, controls are often built into the drier so that it can govern the air inlet temperature. Rate of drying (and therefore time taken) will also be governed by the relative humidity of the air at inlet, within, and at the outlet, and the initial and pick-up of moisture. The humidity in turn will be determined by the rate of flow of hot air that is used. As the drying in the later stages can be relatively slow, partial recirculation of the exiting hot air can then be practised, which will substantially increase the thermal efficiency of the drier. Similarly, heat exchange can usefully be practised.

Several different types of equipment are used and available, as described in the following sections covering static, rotary, horizontal and vertical types. First, the means of providing the hot air (or gases) is important, and depends on the combustible material available economically. Solid fuels can be used, e.g. wood, and the dried outer coverings of the coffee beans, such as husks (see page 22) from dry processing; but also the more convenient liquid fuels and steam. A further factor is whether the hot air actually contacting the drying coffee is provided indirectly through a heat exchanger, or directly, that is, admixed with combustion gases, or indeed smoke-laden. The latter is generally regarded as undesirable for

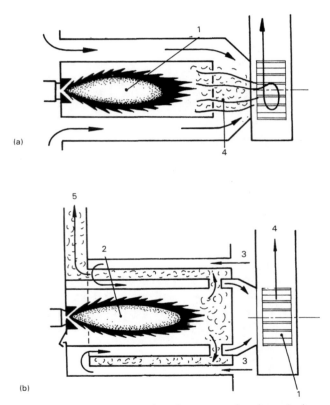

Fig. 2. Typical furnace construction for generating hot air (gases), either (a) directly, or (b) indirectly; showing (1) fan, (2) combustion chamber, (3) cold air inlet, (4) hot air, or mixed gases/hot air exit and (5) cooled gas from furnace.

coffee quality, but admixture with combustion gases can be perfectly acceptable when the combustion itself is efficient. Typical combustion chambers or furnaces for liquid fuels, incorporating a fan for discharging the required hot air (both direct and indirect), are illustrated in Fig. 2.

Further detailed descriptions of drying equipment for coffee berries and parchment are provided by Sivetz and Desrosier;[3] and parameters in drying are discussed by Clarke[4] and also Sivetz and Desrosier.[3] Drying can be carried out either batchwise or continuously, according to the type of equipment.

2.2.1. Static Drier
This type of drier is often very rustic in character, and since it does not require any moving parts is frequently made by the grower himself. The drier will comprise a tray made of perforated sheet metal placed about a metre above a heat–smoke multi-tube exchanger, fired by solid fuel. Clearly, little financial investment is involved, but this kind of drier suffers from lack of flexibility in temperature regulation and a risk of fire from falling debris, and needs constant supervision including hand-raking.

2.2.2. Static Drier with Mechanical Raking
In this drier, a rotary rake mixes continuously the mass of coffee berries, resting on a perforated sheet-metal circular platform through which hot air is blown up from below from a suitable heat generator.

2.2.3. Rotary Drum Drier
Rotary drum driers are still widely used in spite of their low thermal efficiency, since no recirculation or heat exchange from exit air is practised. Different types are available, e.g. the Guardiola drier, though generally more often used for drying parchment coffee; the Torres drier, particularly used in Brazil; and the Okrassa drier. They consist of a rotating perforated drum that holds the coffee and allows the used air to escape. The ingoing hot air is transported through the hollow central horizontal shaft, to which perforated radial tubes are attached. Metal wing plates fixed to longitudinal partitions help mix the coffee. The rotation speed of the drum is about 2–4 rpm. The Guardiola type is illustrated in Fig. 3, a model that has four compartments, which can be used for simultaneously drying small batches of coffee at different moisture content. Especially with cherry coffee of uneven moisture content, it may be necessary to interrupt drying and allow equalisation of moisture by holding in separate bins.

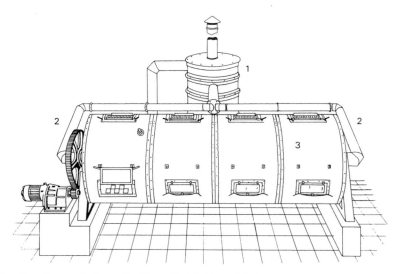

Fig. 3. External view of a 'Guardiola' drier with four compartments; showing (1) hot air generator, (2) hot air ducts to trunnions of (3) rotating drum with charging/discharging ports and exits for the leaving air.

2.2.4. Vertical Driers

There are a number of driers of this type in which cherry or parchment coffee is allowed to fall within vertical chambers while in contact with hot air. Sivetz and Desrosier[3] describe the Moreira drier used in Brazil for drying cherry coffee with hot air provided directly by efficiently burning wood. The so-called American vertical grain drier has been adopted for coffee; Fig. 4 shows a similar design. Such driers are often operated batchwise with recirculation of the coffee until dry and enabling self-equalisation of moisture. Continuous driers are also available incorporating improved internal means of contacting hot air with the coffee.

3. WET PROCESSING METHOD

This technique of processing ensures a much higher quality product. It is always used for preparing mild arabica coffee, particularly in Central and Latin America. It requires processing equipment (for the initial cleaning/classification–pulping–fermentation–washing and drying), an abundant supply of clean water, and the harvesting of ripe fruits only.

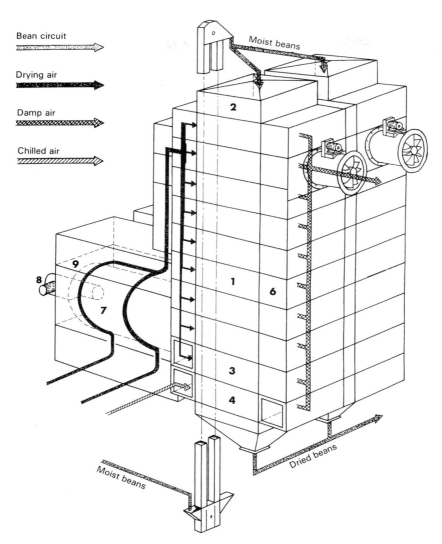

Fig. 4. Construction of a vertical grain dryer, showing flow of moist and dried beans, inlet drying air, exit damp air and chilled air; (1) cells for drying, (2) holding zone, (3) cooling zone, (4) emptying hopper, (5) hot air box, (6) cold air box, (7) combustion chamber, (8) hot air generator and (9) insulated bend.

An important difference from the dry processing method is that most of the outer coverings of the coffee bean are removed before drying.

3.1. Receiving

It is important that only ripe fruits are harvested, as existing pulpers cannot deal with green berries or berries that have dried on the tree. The crop must be transported as quickly as possible to the processing centre to avoid heating of the mass, which can result in irreparable damage (production of discoloured beans, stinker beans, etc.). The crop is generally unloaded into a receiving tank ($1\cdot5$–$2\,m^3$ for a ton of berries) equipped with an overflow weir to remove the floaters. The tank has a syphon ($P > 75$ mm), which transports the heavy berries to the pulper. Such tanks are illustrated in Fig. 5. This simple system fulfils several functions: separating floating from non-floating berries, eliminating sand and stones, and removing leaves/twigs, etc. The floaters, made up of berries dried on the tree and insect-attacked (by *Antestia*, *Stephanoderes*, etc.), are then generally processed by the dry method.

3.2. Pulping

Pulping consists of removing the exocarp (outer skin) and the major part of the mesocarp (fleshy portion), which is referred to as the 'pulp'. It is a delicate operation, for if the bean itself is damaged it will be susceptible to microbial attack and penetration by undesirable substances. The principle on which the pulpers work is essentially the 'tearing off' of the exocarp and mesocarp of the berries, which operation is carried out under running water. Conventional pulping does not remove all of the mesocarp, which is the mucilage adhering to the parchment surrounding the actual coffee beans together with the testa. The yield expressed as dry green coffee varies with the species: arabica at 15–19%, robusta at 18–23% and arabusta at 13–16%. The various types of pulper are described in the following sections.

3.2.1. Disc Pulpers

The berries pass through a feed-roll assembly and are advanced by means of discs (with a bulbed surface) generally 18 in (46 cm) diameter fixed to a horizontal shaft rotating at 120 rpm. The berries are stripped on contact with the lateral pulping bars or the breastplate. The beans are then separated from the pulp by a plate with a straight sharp edge fixed at right angles to each disc, which allows the pulp to pass but retains the beans. Such a pulper is illustrated in Figs 6 and 7. The pulping bars and

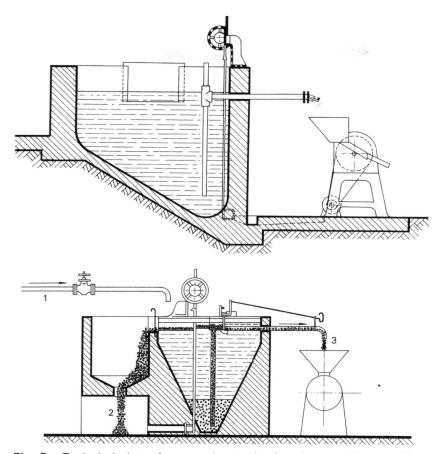

Fig. 5. Typical designs for reception tanks for cherry coffee providing floating and cleaning facility; showing (1) water supply, (2) outlet for floating cherry, (3) siphon outlet for heavy coffee to pulper.

plates can be adjusted in position, according to the diameter of the berries. For arabica coffee, a space of 2 mm between disc and plate and of approximately 5 mm between disc and pulping bar would be a starting position. The discs can be made of rough-surfaced cast iron with bulb projections, or smooth and covered with flanges made of punched copper. A typical flow rate is approximately one metric ton of berries per hour and per disc, with a nominal running speed of 120 rpm. The horsepower required would be 1·0 for a single disc and 2·5 for four discs.

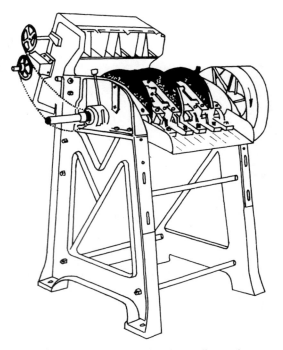

Fig. 6. Outside view of a three-disc pulper.

3.2.2. Drum Pulpers

The design of this type of pulper dates back some years, but is still in use, with variants. It comprises a rotating drum made of punched copper sheet with projecting mounds each having a cutting edge, and an adjustable breastplate with a smooth, channelled or slotted surface, the sectional area of which decreases to cause a squeezing action on the coffee as it passes through. The distance between the breast and drum is adjustable, usually by tensioning springs. There is also a plate placed against the drum, which has the same separating function as in the disc pulper. Such a pulper is illustrated in Fig. 8. They are less easy to adjust than disc pulpers and are more fragile. A typical drum size would be 15 in (38 cm) diameter with a width of 12 in (30 cm) having a capacity of one metric ton of berries per hour at a nominal speed of 120 rpm and requiring 1·0 hp. The capacity will be doubled by doubling the width and including four sets of slots.

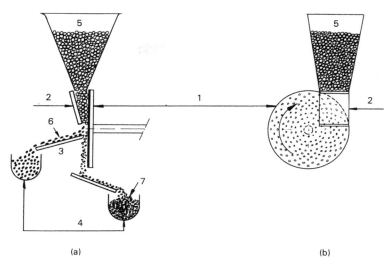

Fig. 7. Diagrammatic details of a disc pulper in (a) side and (b) end sectional view; showing the (1) rotating disc, (2) pulping bar, (3) separating plate, (4) receiving troughs, (5) cherries, (6) beans, (7) pulp.

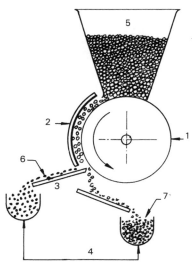

Fig. 8. Diagrammatic details of a drum pulper in side view; showing (1) rotary drum, (2) breastplate, (3) separating plate, (4) receiving troughs, (5) cherries, (6) beans, (7) pulp.

3.2.3. Pulper–Repasser System

The performance of a pulper is never quite perfect on account of the fact that berries have different diameters. If the berries are very heterogeneous in this respect, it is highly recommended to precede pulping by some size grading and to pulp the different grades or categories separately. Generally, therefore, a pulping line comprises a pulper, a sieve and a 'repasser', as illustrated in Fig. 9. This combination allows the pulping bars of the pulper to be set a little less tightly, so that small berries will pass through. These are separated by rejection from the sieve and subsequently pulped by the pulper–repasser, which is either a small drum or disc pulper. It is difficult to recommend suitable diameters for the apertures of the sieves, since each species/crop of coffee requires a preliminary appraisal of the range of cherry size present. Rotary drum or vibrating sieves are used as illustrated in Fig. 10.

Water consumption for all the foregoing operations in the wet process from receiving to pulping is approximately $40 \, m^3$ per metric ton of green coffee or $8 \, m^3$ per metric ton of berries.

3.2.4. Raoeng Pulper

A few years ago, this machine was one of the rare innovations brought to the wet processing line. It is designed to pulp the berries mechanically, but also to remove the mucilage from the coffee, which, as already noted, the previous machines cannot do. This machine consists of a horizontal

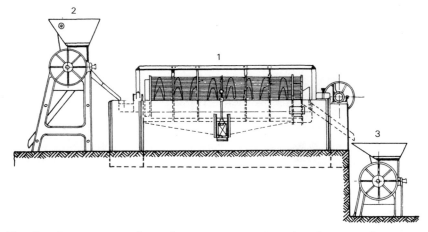

Fig. 9. Arrangement of a pulper–repasser system showing use of a rotary drum screen (1), (2) pulper, (3) repasser.

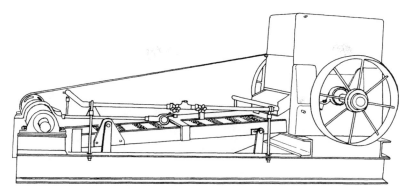

Fig. 10. View showing construction of a vibratory separating screen for use with a repulper.

perforated rotating drum inside a fixed cylinder, both of which are slightly cone-shaped and laterally adjustable within each other. The inner drum is fitted with longitudinal and transverse cleats, which advance the coffee, and pressure is developed against an adjustable floodgate at the outlet. Water under pressure is fed by a hollow shaft to the inside of the drum, where it spurts through the perforations in the drum and passes between the drum and outer jacket. There the mucilage and fine-ground pulp is forced out through small slots in the jacket but the parchment coffee is retained until discharged. This type of pulper is illustrated in Fig. 11. There is only one type and size available at present, the Aquapulpa Major, which has a capacity of 3000 kg of berries per hour at a drum rotation speed of 400 rpm, requiring a horse power of 25.

3.2.5. Roller Pulper
This type of machine has not been developed to any great extent. It works by pressing the berries between two drums, one of which is ribbed to guide the berries to the centre and the other of which is smooth.

3.3. Separation/Classification

The berries are not perfectly sorted in the receiving tank; there is a further opportunity after they have left the pulper, again by flotation by different means. This is particularly appropriate for mild arabica coffee, since, for example, floating parchment coffee (i.e. with the pulp and mucilage removed) contains very few healthy beans. The following methods can be used at this stage.

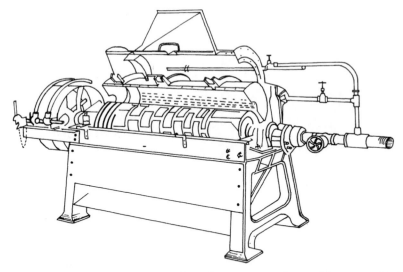

Fig. 11. View showing construction of a Raoeng pulper, with outer cylinder opened to show internal rotating drum.

3.3.1. Draining
Using small vertical boards acting as a spillway for circulation in running water, the lightest beans go the furthest, and the heaviest are stopped by the first boards.

3.3.2. Use of Fermentation Tanks
Fermentation to remove mucilage is described on page 17, together with the operation of the tanks.

3.3.3. Aagaard Densimetric Grader
This grader is based on the jigging principle used to separate minerals from coal. Jigging consists of placing the heterogeneous matter on a sieve submerged in water. The sieve moves to create ascending and descending currents, and with each pulsation the beans are suspended and resettled. The heaviest beans move towards the bottom of the layer and the lightest move towards the top. When this machine is positioned between the pulper and repasser, it separates the mass into three categories: small berries that have not been pulped, heavy beans that are removed by a scraper conveyor, and light beans.

3.4. Fermentation

The object of fermentation is to hydrolyse the mucilage, already mentioned, in order to facilitate its final removal during subsequent washing. If mucilage remains present during drying, there is the risk of undesirable fermentation, which is detrimental to the quality of the coffee. Biochemically speaking, the hydrolysis of pectins is caused by a pectinase already in the fruit, but the reaction is accelerated by different micro-organisms, such as *Saccharomyces*, which also have pectinolytic properties. The rate of hydrolysis depends upon the temperature; consequently, it is necessary to adapt the length of the fermentation period to existing ambient conditions. Some micro-organisms can cause off-flavours to develop, particularly in prolonged fermentation. It is important to prevent the development of harmful species (moulds, aerogenic coli) by encouraging the development of acidogenic species. Control of pH is important to avoid excess formation of acids, such as propionic (Wootton[5]). Fermentation can be carried out either 'dry' or under water. Mixed fermentation means dry fermentation during the initial stages in order to acidify the environment rapidly so as to protect against yeasts and moulds, and then a period of soaking in water. Studies carried out by Wootton[6] in Kenya have shown that the pulped bean loses approximately 1% of its dry matter during fermentation. Some water-soluble substances (polyphenols/diterpenes) undergo exosmosis, and their partial elimination improves the quality of the coffee by reducing brew hardness and bitterness. Soaking in water therefore refines the bean. It appears that pectinolysis does not have any effect on final characteristics. It would therefore be perfectly feasible to use mechanical or chemical means of removing the mucilage, followed by refining in clean, acid water.

Fermentation is most often carried out in concrete tanks, which vary considerably in size. A system of grids and vertical liftgates (usually made of bronze or cast iron) allows the discharge of either water alone, or the water and the coffee. Efforts have been made to improve fermentation conditions, and in particular to facilitate stirring/mixing. In an experimental factory at Soubre in the Ivory Coast, fermentation was arranged to take place in vats, using a process inspired by that of vinification. Stirring is carried out by an air-lift system, and the floating pulped berries are easily removed. Only one manual operation is involved—opening the sluice gates—but complete automation is possible.

The length of the fermentation time differs mainly as a function of climatic conditions and the condition of the crop. In regions of low altitude, where robusta coffees are grown, fermentation time is very short

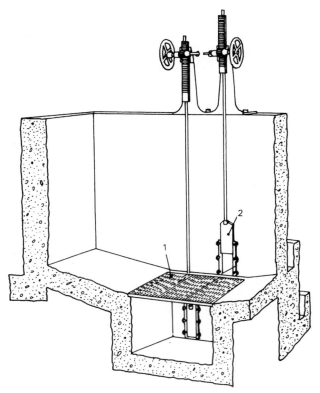

Fig. 12. Constructional details of a typical fermentation tank, showing (1) grid and (2) sluicegate.

(e.g. 16 h); whereas at high altitudes, fermentation generally takes some 48 h. Figure 12 illustrates the typical construction of a fermentation tank.

Beans that have been damaged during pulping can develop tainted to foul flavours together with a 'foxy' appearance if the water used has been too alkaline or contains too much organic matter.

3.5. Washing
All mucilage has to be removed by washing by one of several different methods so as to prevent any subsequent proliferation of micro-organisms.

3.5.1. Manual Washing
Washing is frequently carried out manually in the fermentation vats themselves. If, however, the depth of the mass is too great (say 0·6–0·7 m),

the job becomes difficult. The wet parchment coffee may also be washed in drains (say 60 m long) with a gradient, being moved along by paddles, which may also provide some densimetric grading (see page 16). Washing and gravity grading will require some 25–30 m^3 of water per metric ton of green coffee.

3.5.2. Mechanised Washing

Three different types of washer are of interest. The batch horizontal washer has a perforated metal drum with internal agitator blades rotating at some 90 rpm. The base of the drum is submerged in a tank of water. The vertical washer has rotating blades (18 rpm) inside a vertical vat, as illustrated in Fig. 13. Initially, no water is used to facilitate mechanical removal of the mucilage. A centrifugal pump with profiled blades is also used, in particular the Kivu pump, which is well suited for washing by energetic stirring and conveying a mixture of the coffee mass and water at a volumetric ratio of 40:60 with a flow rate of 20–70 m^3 per hour. The pump needs to be primed, but with 7·5 hp will provide a lift of 9 m.

3.6. Draining and Pre-drying

In this stage, the moisture content of the mass, or wet parchment coffee, is reduced from 60 to 53%, which is fairly easily done, in either a draining tower or on trays. Alternatively, the coffee mass can be left in the sun for two or three days and stirred frequently, when the moisture content can reach 45%.

The total arrangement of the wet processing system from reception to draining is shown in Fig. 14, which also illustrates the use of downward gravity movement wherever possible.

3.7. Drying of the Parchment Coffee

As already discussed, drying is always a delicate operation in the processing of coffee. If not carefully carried out, there can be a disastrous effect on quality as regards both appearance and brew characteristics.

Drying of coffee comprises a so-called wet stage, ranging from 60 to 30% moisture content, and a hygroscopic stage, below 30%. Gibson[7] distinguishes two additional intermediate stages in parchment coffee: the 'white' stage, when the moisture content is reduced to 30%, but which must be carried out slowly in the shade to prevent cracking or breaking of the parchment cover of the beans; and the 'soft' black stage, where the moisture content is reduced from 30 to 23% (as-is figures), but which must be carried out in the presence of sunlight (that is, natural drying).

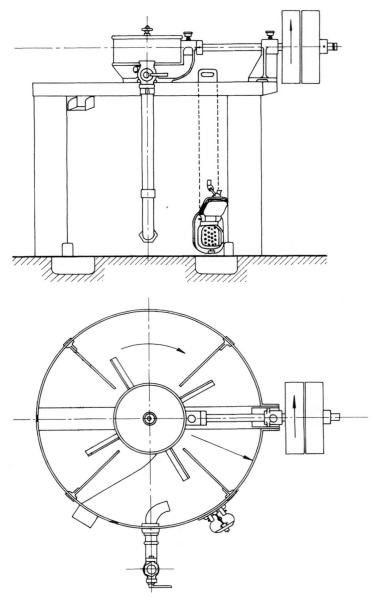

Fig. 13. Constructional details of a vertical washer for fermented and pulped parchment coffee.

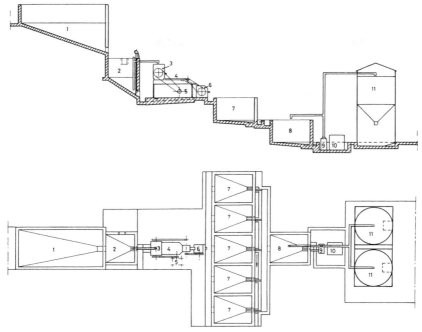

Fig. 14. Diagram of a wet processing system; showing (1) receiving tank, (2) syphon, (3) depulper, (4) separating screen, (5 and 10) 'heat engine', (6) repasser, (7) fermentation tanks, (8) washing tanks, (9) thick matter pump and (11) drainage hopper.

According to this worker, artificial drying can be carried out without danger at the start to remove lightly bound water, and again at the end of the drying period when the final required moisture content of 11% is being achieved, for subsequent satisfactory storage stability. During the wet-stage drying, enzymatic changes can occur, and during the hygroscopic stage, chemical transformations can take place at high temperatures. Cracks and breakages in the parchment, which can arise from mechanical movement and other causes, are known to be undesirable for quality in sun-drying.

Sun-drying is the most widely used method. In Kenya, great care is taken to ensure that this operation is successful. Drying is usually carried out on trays, with coverings that can be used during the hottest hours of the day. With the coffee spread out at a level of $10-15\,kg\,m^2$, the drying rate will be between 10 and 15 days. Drying on solid surfaces such as concrete is not recommended, as cracks in the surface provide centres for proliferation of micro-organisms. Artificial drying is also largely used, since it is becoming more and more difficult for large estates to dry coffee naturally. Again, care has to be taken with the drying conditions, especially with the higher qualities; for example, in the hygroscopic stage, the temperature must not go above 55°C.

4. CURING

The operations carried out subsequently to those of section 2 (for dry processed coffee) and section 3 (for wet processed coffee) are collectively known as curing, and most often are performed centrally, away from the growing and other processing areas. In this way, the coffee is prepared in the green bean condition for consumption and export.

4.1. Redrying
Coffee, whether wet or dry processed, arriving from farms and estates is not always sufficiently dry, and it is often necessary to dry certain batches further. Apart from providing storage stability, a moisture content of 11% enables husk and parchment to be removed more easily.

4.2. Cleaning
Dry parchment coffee, and particularly dried coffee cherry, may contain many impurities, e.g. pebbles, pieces of metal and other foreign bodies. These need to be removed by a preliminary cleaning to protect the curing equipment. Cleaning can be carried out by use of a hopper with screens to remove large and medium-sized impurities, followed by a magnetic separator to remove metal pieces and a cleaner–separator which combines the effect of sifting and pneumatic dust removal. An air-float separator consists of an inclined flat-bed screening unit, up and across which a current of air is blown to cause fluidisation. With a reciprocating movement of the bed, the most dense particles circulating at the bottom rebound successively higher up the inclined bed. This system essentially sorts particles with the same diameter but different densities.

4.3. Hulling
In wet processed coffee, the hull or dried parchment layer immediately surrounding the beans has to be removed. In dry processed coffee, the whole of the dried outer coverings of the original cherries or husks have

to be removed. In either case, hulling (or, strictly speaking, dehusking or decortication in the case of dry processed coffee) may or may not remove a final layer closest to the beans, now called the silverskin, deriving from the testa. Some equipment types may be used for both types of processed coffee with modifications for each. If removal of silverskin is required, it may require separate equipment following hulling, called polishers.

4.3.1. The 'Africa' Huller

This type of huller is made by a number of different manufacturers, e.g. Gordon, Kaack, McKinnon and Bentall and is widely used. It is a simple and robust machine based on the original Engelberg patent, and is illustrated in Fig. 15. It comprises a steel screw, the pitch of which increases as it approaches the outlet; a rotor that turns inside a closed horizontal cylinder with a perforated grid at the bottom; and a knife fixed to the outer frame parallel with the rotor shaft. Husks or parchment debris are sucked through the grid by a fan. Capacities are indicated in Table 3.

Table 3
Typical capacities of 'Africa' hullers for different models

Model No.	Flow rate for:		Power (hp)	Rotor speed (rpm)
	Parchment	Husks		
	(kg per hour)			
0	—	1000	30–35	—
1	650	440	10–12	400
$1\frac{1}{2}$	410	280	8–10	400
2	320	220	6–8	450
5	170	110	4	550

4.3.2. The Smout Huller–Polisher

This huller uses a rotating screw with a helical pitch, increasing towards the discharge end, which turns inside a horizontal casing with two matching concave covers with spiral grooves. The spirals of the rotor and casing turn in opposite directions. The rotor pushes the coffee forward, while the parchment/husk is released by friction, and the pressure is adjustable by a balanced floodgate at the outlet to regulate the flow. The machine is also equipped with an aspirator fan to assist removal of debris. Table 4 indicates the capacities available. There is considerable inertia at

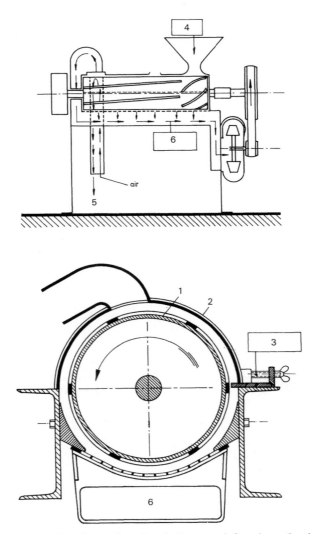

Fig. 15. Constructional details of a huller used for decortication, based upon the Engelberg principle; showing (1) rotating inner screw, (2) outer cover with grid, (3) knife, with position adjustment, (4) entry port for the dried coffee cherries, (5) exit port for the dehusked coffee and (6) passageways for husks and dust.

Table 4
Typical capacities of Smout hullers for different models

Smout machine No.	Casing length (in)	Flow rate (parchments) (kg per hour)	Power (hp)	Rotor speed (rpm)
1	54	2000	32	60
$1\frac{1}{2}$	42	1000	16	100
2	36	400	7	120–140
$2\frac{1}{2}$	28	200	5	140
3	18	140	3	120–140

start-up, which justifies the inclusion of an electric motor with a powerful starting torque. Although these machines function well, there is considerable wear.

4.3.3. The Okrassa Huller–Polisher

This machine has two compartments, the first resembling that of the Smout huller. The second is specifically designed for polishing, and is illustrated in Fig. 16. Models have capacities ranging from 1000 to 2000 kg of parchment per hour.

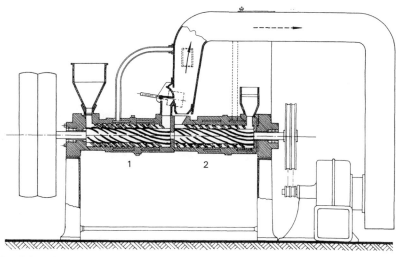

Fig. 16. Constructional details of the Okrassa huller–polisher, showing the two separate compartments: (1) huller, (2) polisher.

4.3.4. The 'Descascador' Huller

This machine, of Brazilian origin, comprises a cage made of perforated sheet metal in which a centred blade with knives rotates. The dried coverings of the coffee are removed by friction between the cage and the knives, while the hulled beans pass through the perforated breastpiece. This huller is also equipped with a rotary green coffee cleaner, which retrieves unhulled coffee. This huller provides excellent results, with few beans being broken and with none of the overheating effects on the beans that can be characteristic of other types. There are a number of similar machines on the market, with flow rates ranging from 150 to 9000 kg h^{-1}.

4.4. Size Grading

After the outer coverings have been removed, it is advantageous to size-grade the green coffee beans of whatever kind. A higher price can be obtained for the larger beans; caracoli (peaberries) can be separated out from normal 'flat' beans; subsequent gravimetric sorting is easier; and finally the roaster can roast more evenly.

Two types of size grader, based on the sieving principle, are available, one using screens mounted in drums, or a reel grader, as illustrated in Fig. 17, and the other a vibrated flat bed of rectangular shape. Apertures of the screens are either round with diameters at increments of 1/64 in, or oblong for peaberries.

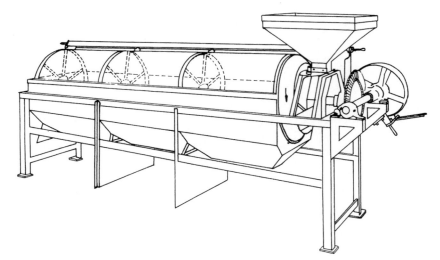

Fig. 17. Bean size grader with rotating drum; screens not shown.

4.5. Density Sorting
Green coffee at this stage will still include dust and other light particles, together with beans of light density, or which are deformed, discoloured or insect-attacked. A number of different sorting methods are used for progressive separation of undesired beans and other matter.

4.5.1. Pneumatic Sorting
Sorting by this method can be done in one of two ways: the use of an upward current of air, or by a gravimetric table. Pneumatic sorting depends on the use of two physical parameters of the bean, i.e. density and diameter. For this reason, pre-size separation is desirable, since two spheres with different diameters can in fact have the same settling velocity when their density is different, which would not enable any separation. The fundamentals of this separation are apparent from the following basic equations.

Settling velocity of a particle in a fluid (such as gas or air) is given by

$$v = K\sqrt{D_s(\rho_s - \rho_{fl})}$$

where D_s is the diameter of the particle, ρ_s its density and ρ_{fl} that of the fluid. For two solid particles (e.g. coffee beans of two different diameters, D_{s1} and D_{s2} and densities of ρ_{s1} and ρ_{s2}), when $D_{s1} < D_{s2}$ and $\rho_{s1} < \rho_{s2}$ for equal settling velocities, then

$$\sqrt{D_{s1}(\rho_{s1} - \rho_{fl})} = \sqrt{D_{s2}(\rho_{s2} - \rho_{fl})}$$

from which it follows that

$$\frac{D_{s1}}{D_{s2}} = \frac{\rho_{s2} - \rho_{fl}}{\rho_{s1} - \rho_{fl}} \approx \frac{\rho_{s2}}{\rho_{s1}}$$

If D_{s1} were 17/64 in, and $\rho_{fl} = 1.20$, and if D_{s2} were 16/64 in, and $\rho_{s2} = 1.27$, there would be no separation.

The 'Catador' is based on the principle of pneumatic separation by the use of an upward current of air created by a radial blade fan rotating at about 550 rpm within a vertical chamber (see Fig. 18). The Duplex model (Fig. 19) can separate normal beans into two categories, heavy and medium heavy, from other material. Even though pre-size grading has been practised, in reality separation will be imperfect for a number of reasons, i.e. the drag coefficient of the beans is variable due to their not being perfect spheres, and in turbulence a bean may present itself differently in the flow of air; the velocity of the air is lower closer to the

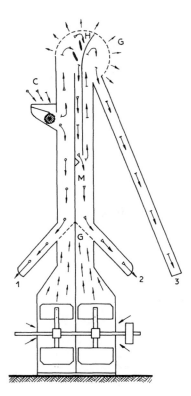

Fig. 18. Diagrammatic cross-section of a 'Catador' for pneumatic sorting, showing (1) whole product, (2) middle product, (3) splits and trash, (C) inlet, (G) grid, (M) block, (H) head.

walls; the adjustment mechanism can cause deflection of the air flow; and the velocity and flow rate of the air are affected by the resistance provided by the surface of the bean, thus a variation in the flow causes a variation in the classification limit between light and heavy beans.

4.5.2. Gravimetric Sorting
In the gravimetric separator, the principle is similar in that careful size grading should be carried out beforehand, as the fluidising velocity depends upon the size of the beans as given by an equation similar to that given previously.

4.6. Colorimetric Sorting
Discoloured beans can be sorted out by hand together with removal of more obvious visual defects, but electronic sorters have made great progress since 1959. Black beans in particular need to be removed.

GREEN COFFEE PROCESSING

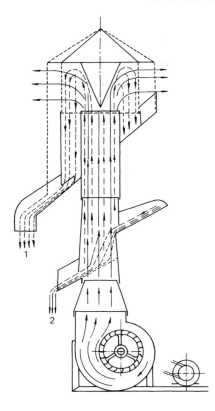

Fig. 19. Diagrammatic cross-sectional view of a Duplex model pneumatic sorter, showing separation into light (1) and heavy (2) fractions.

4.6.1. Manual Sorting

Sorting by hand is of course extremely tedious and labour-intensive, but some 80 beans can be removed per minute when using both hands. Stationary tables can be used for the sorting, but more usually belt conveyors manned by people on both sides of the belt are employed.

4.6.2. Electronic Sorting

In these machines, beans are distributed by slide, belt or rollers, and pass one by one past electronic eyes assessing colour, which then control a mechanical ejector for the removal of the beans as required. The sorters use either monochromatic or bichromatic light (to increase their sorting capability) and can be adjusted to eliminate either more or fewer defective beans. The flow rate of the beans is of the order of 100–150 kg per hour per trough, dependent also upon the size of the beans. The construction of a typical 'Sortex' machine is illustrated in Fig. 20.

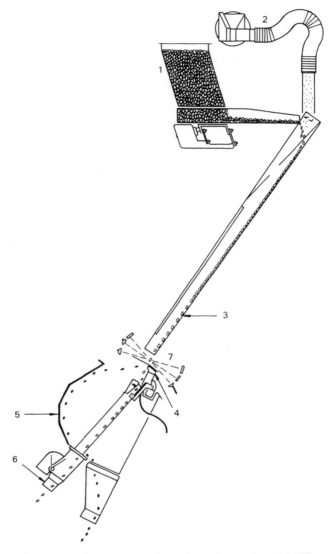

Fig. 20. Diagrammatic representation of an electronic sorter (Sortex model) showing (1) hopper with vibrator, (2) dust extraction, (3) slide chute for incoming beans, (4) defect bean ejector, (5) receiver for rejects, (6) collector for acceptable beans, (7) optical sensing area.

Adjustment will depend upon the number and importance of the defects in the coffee, but also on the profitability of the operation. Clearly, losses by removal must be compensated by the incremental value of the sorted product, as by the following equations from Wilbaux.[8] It is important to judge the advisable degree of thoroughness of electronic sorting, which can be done by estimation from laboratory tests.

If x is the percentage amount of waste eliminated (e.g. defective beans); c is the cost of sorting per kilogram of unsorted coffee; P is the ex-factory cost of unsorted coffee; and p is the incremental value of the sorted coffee per kilogram, then for 100 kg of unsorted coffee, the following must apply:

$$(100 - x)(P + p) \geq 100(P + c)$$

If the waste also has a value of P' per kilogram, then

$$(100 - x)(P + p) + x \cdot P' \geq 100(P + c)$$

The possible types of defect that can be eliminated by electronic sorting are, for monochromatic machines, black beans, grey or dull beans, some beans attacked by insects (*Antestia*); for bichromatic machines, black, grey, brown, foxy, unripe green, white or marbled beans, some beans attacked by insects, and dried cherries.

5. STORAGE

Storage of coffee in a producing country can be in the form of dried cherry, or dry parchment coffee, or, of course, cured green coffee. Storage conditions need not be exactly the same, since both dry husk and parchment provide a good protection against insects and also a barrier against moisture transfer.

As already mentioned, the moisture content of coffee for satisfactory storage should not be over 11% w/w. At this level, mould growth and enzymatic activity is minimal. When the relative humidity of the air is over 74% (corresponding to an equilibrium moisture content in the coffee of about 13% w/w, dependent upon temperature, see Chapter 2 in this volume, and also in Volume 1), certain innocuous moulds develop, i.e. *Aspergillus niger, A. ochraceous, Rhizopus* sp. A relative humidity of more than 85% is required for the multiplication of yeasts and bacteria (corresponding to about 18% w/w in the coffee). Dependent upon the relative humidity/temperature of the air and the length of time in storage, so (bio-)chemical changes take place in the coffee, resulting in discoloration (often a bleached or white appearance). However, the actual chemical changes are not yet fully understood.

Among the insects capable of causing spoilage (e.g. by boring holes) in green coffee during storage are *Araecerus fasciculatus* (the most dangerous), *Lasioderma serricornis*, *Tribolium castaneum* and *Carpophilus* sp.

A traditional storage building is suitable for storing green coffee in sacks, as long as it is heat-insulated to prevent marked variations in temperature and perhaps also ventilated. Sacks are stored on pallets placed some 0·5 m minimum distance from walls. The building should be completely clean and have smooth walls, and all openings should be equipped with devices to prevent insects from entering.

Coffee can also be stored in bulk, as is also done for buffer stocks in a roastery in the consuming countries. Metal or concrete storage bins should, however, be equipped with anti-breakage devices such as chutes, to prevent bean breakage. Large-capacity silos are rarely used, though there is a good example in Colombia for the storage of parchment coffee (Medellins). Some 60 000 metric tons of dried coffee cherry can be stored in bulk at the Seric factory in Tombokro in the Ivory Coast.[9] The storage building measures some 400 m by 50 m, and is made of glued, laminated wood. A chain conveyor is used for filling, and shovel loaders for emptying.

Alternative systems of storage have been studied, e.g. vacuum and inert-gas storage in containers. Current results are promising for vacuum storage, which has the advantage of solving the problem of insects and mould growth. Disposable plastic containers of 25–1000 litres are envisaged, which can be stored in the open air and do not require the cost of a building. Otherwise, insects have to be destroyed, and the most widely used insecticide is methyl bromide which has the disadvantage of being dangerous. It is generally applied under tarpaulins outside the building, at the level of $15\,\mathrm{g\,m^{-3}}$.

6. HANDLING

Coffee needs to be transported at different times; belt conveyors, bucket elevators and screw conveyors are all used, as is pneumatic conveying. For green coffee, the latter has to be carefully applied by avoiding bends in the pipework and high velocities. Robusta coffees tend to be more brittle than arabica, and the percentage of broken beans will be found to increase with coffee moisture contents below 10% w/w. Conveying to silos, as already mentioned, can be hazardous unless suitable devices to prevent bean breakage are included.

REFERENCES

1. Medcalf, J-C., Lott, W. L., Teeter, P. B. and Quinn, L. R., *Experimental Programs in Brazil*, IBEC, Research Institute, New York, 1955, 30.
2. Roelofsen, P. A., Archief. Koffiecultuur Ned. Indïe, 1939, **XIII**, 151.
3. Sivetz, M. and Desrosier, N. W., *Coffee Technology*, AVI Westport, Conn., 1979.
4. Clarke, R. J., in *Coffee: Botany, Biochemistry and Production of Beans and Beverage*, Eds M. N. Clifford and K. C. Willson, Croom Helm, London, 1985.
5. Wootton, A. E., *Proc. 2nd Coll. ASIC*, 1965, 247–58.
6. Wootton, A. E., *Proc. 5th Coll. ASIC*, 1967, 316–24.
7. Gibson, A., *Proc. 5th Coll. ASIC*, 1971, 246–54.
8. Wilbaux, R., *Technologie du café Arabica et Robusta*, Publication de la Direction de l'Agriculture, des Forêts et de l'Elevage, Brussels, 1956.
9. Richard, M., *Proc. 8th Coll. ASIC*, 1977, 187–96.

Chapter 2

Grading, Storage, Pre-treatments and Blending

R. J. CLARKE*

Formerly of General Foods Ltd, Banbury, Oxon, UK

1. INTRODUCTION

Following the processing of coffee cherries to provide the green coffee beans of commerce, described in the first chapter, the next stages of processing, carried out only in the consuming countries, involve the conversion of these beans to coffee products in forms suitable for brewing to give the final beverage. The key operation is the roasting of the beans (described in Chapter 4), but the green coffees that are available require first to be selected and then in most cases blended before roasting. The green coffees may also need to be stored before use, and furthermore may need to be pre-treated by certain physical operations for cleaning, or indeed for removal of some particular component, such as caffeine, which may be unwanted in certain markets. All these different operations (except decaffeination, which is covered in Chapter 3), are therefore described in the following sections.

2. MARKETED GRADES

As already clear from Volume 1 (see Chapter 1), the number of different types of green coffee available at different times on the market is very large. Each producing country is generally able to provide a range of qualities. Different green coffees are therefore broadly (or in some cases

* Present address: Ashby Cottage, Donnington, Chichester, Sussex, UK.

in detail) characterised by specifications provided by marketing authorities in the various countries of origin.

2.1. Systems of Specification

Specification will or should cover the items of information about a particular green coffee shipment shown in Table 1. Some of the information can only be derived after appropriate sampling.

The actual use in practice of such systems of specification differs from country to country. Their origin is to be found in the early system set up by the New York Sugar and Coffee Exchange to deal with the importation of green coffees from Brazil and Central/Latin America (i.e. milds and other wet processed arabicas). Similar systems are, however, a feature of

Table 1
Characteristics of green coffee specifications

Item	Type of description
Designation	Species (arabica, robusta or other commercial).
	Processing ('wet' or 'dry'; whether 'washed and cleaned' for dry process robusta; or 'polished' for wet process arabica).
	Geographical origin (name of country, local area of production, height at which grown, port of shipment).
	Crop year
Quality factors	
Liquoring/flavour	Word description
Bean size and shape	Size characterised by grade letters or numbers or word terms by reference to results of screen analysis of sample
Defective beans	Quantity characterised by type numbers or word description, based on physical counting or weighing from a sample
Extraneous matter (coffee and/or non-coffee)	Either characterised separately, or in conjunction with defective beans (as defects) in type numbers/description
Colour	Word description
Roasting characteristics	Word description of appearance/uniformity, defect counting
Bulk density	Numerical value from measurement
Overall	Word description or grades/standards, combining all or some of above factors

the specifications of most producing countries (and some other importing countries, e.g. London Terminal Markets for robusta and arabica). It should be emphasised that not all the items will necessarily be considered, nor appear in actual contracts between seller and buyer. Minimally, however, country of origin, species, type of green processing and usually bean size grade will be clearly established, and such information stencilled on bags together with net weight.

Details of specifications for marketable coffees from different producing countries are available through the various marketing channels. These specifications have changed relatively little over many years. They have also been described in various other publications and texts; reference should be made to Jobin,[1] who gives very full and up-to-date details from all countries, and others,[2,3] and to occasional articles in *World Coffee and Tea*.

Many items of specification may indeed not be explicit, since these can be covered by 'according to sample'. Proper sampling is central to specifications, and takes place at the port of entry and before export. Different sampling procedures are used to obtain what is desired—a representative sample from a consignment. An ISO Standard[4] attempts to harmonise procedures; it suggests sampling from at least 10% of the bags to provide at least three samples of 300 g each, though often in practice the level of sampling may well be higher. A further ISO Standard[5] sets out details of a recommended type of bag sampler. The third sample (completely sealed) can be used for arbitration purposes in cases of dispute, as discussed by Marshall.[6] See also Volume 1, Chapter 1.

It will be noted that a number of the variable compositional characteristics of green coffee are not usually subject to specification, in particular moisture content. In practice, moisture contents will not exceed about 13% w/w, since otherwise deterioration will (or is likely to) occur. Some changes in moisture content may occur in shipment, though would be detectable by change of bag weight, for which—with the exception of minor losses—monetary compensation can be contractually arranged (together with non-moisture losses by spillage, etc.). The measurement of moisture content in green coffee is a somewhat complex problem (see Volume 1, Chapter 2). In any event, it is essential that the sample for moisture measurement be placed in a moisture-proof package or container immediately after sampling. Caffeine content is also not a matter for specification (except for decaffeinated green coffees, when it may also be desirable to know the moisture content), though range values are inherent in the species of the coffee. Compositional factors are more important in the roasted coffees themselves, and depend upon the degree of roasting.

The quality factors set out in Table 1 are discussed individually, with particular regard to the different practices of producing countries, in the following sections. Many of these factors are determinable by test methods published by the International Standards Organization. There is relatively little national legislation on these factors, nor need there be,[3] though any such legislation should be consulted for each country.

2.2. Liquoring/Flavour Characteristics

Flavour is the main and most important criterion of coffee quality, that is, of the consumable brew prepared after roasting, grinding and infusion with hot water (so-called 'liquoring' or 'cupping'). Coffee flavour will also be a function, especially in consumer reaction, of the level of roasting applied (see Chapter 4, section 2), of grinding (see Chapter 4, section 4) and of the type of brewing appliance used (see Chapter 8). For the initial assessment of flavour or so-called cup quality, by both buyers and sellers, the brewing method of 'steeping' is almost universally used (e.g. some 10 g of roast and ground coffee per 150 ml boiling water left for 5 min), generally with a medium roast coffee prepared in a sample roaster and with a relatively fine grind. Exact details of procedure will differ from one location to another, and a proposed draft International Standard seeks to harmonise these.

This factor is not one very amenable to description for a specification; there is, however, an expectation of flavour quality from each of the main groups of marketed coffees, e.g. as divided by the International Coffee Organization in their quota system; and, of course, for more specifically designated coffees. Washed arabicas (or milds) from Kenya, Tanzania and Colombia are especially expected to have a 'fine acidity with aromatics'; whereas dry processed robustas are expected to be neutral (though 'straw like') with varying degrees of harshness, and Brazils are expected to be somewhat intermediate. Brazils, however, are officially categorised by reference to a flavour scale (strictly soft–soft–softish–hard–Rioysh–Rio) reflected in degree of absence of Rio or medicine-like (iodoform) flavour, associated with certain areas of production. In general, it is the off-flavours that are more important for downgrading, with various identifying names[7] both in wet processed arabica and in robusta (see also section 2.4).

In many countries, overall quality terms are often used; thus 'usual good quality' (UGQ) in Colombia, or 'fair average quality' (FAQ) with robusta coffees. These statements are implicitly combining cup quality assessment with other observable characteristics of the coffee, such as numbers of defects within a defined group. In Kenya, where individual consignments of coffee are placed in auction, the results of official liquoring tests are combined with examination of other characteristics, i.e. green and roast bean appearance, to provide a range of numbered standards for quality (from 1 to 10, the poorest (with >10) rejected for export) for buyer guidance only. Quality connotations are also implied but are also real, especially with Central American arabicas, by reference in the designation to the height at which the green coffee is grown and processed; e.g. Costa Rica, low-grown and high-grown Atlantic; and similarly for Nicaragua and El Salvador. Guatemala uses designations of strictly hard, hard and semi-hard, for its better qualities, referring to the height at which grown and not, of course, to Brazilian usage of the term 'hard'. There will also be a general relationship of flavour quality to bean appearance, size and number of defects. Wet processed arabicas (especially from Kenya and Colombia) will carry relatively few defective beans; but dry processed coffees, whether arabica or robusta, will often contain many more, so these defects are separately quantified and specified.

2.3. Bean Size and Shape

As already described in Chapter 1, green beans are graded for size by large-scale screening or other machinery and marketed in specific size grades. These size grades relate usually to the results of laboratory screen analyses (rather than the aperture sizes of the large-scale screens that have been used); so that particular grades may simply be described as large, medium or small (or in more detailed categories, such as bold in Brazil), or by letters (A, AA, etc., as in Kenya), or by numbers (grades 1, 2, 3 as in the Ivory Coast and other OAMCAF countries). The need for a range of size specifications is more usual for dry processed coffees than for wet, so that in Colombia there is really only one size grade, 'Excelso' (though larger beans are known as 'Supremo').

Arabica green coffee beans, especially wet processed, tend to be flat-shaped (so-called 'flats', one surface flat with a central cleft), though they may contain so-called peaberries (French *caracoles*; Spanish *caracoli*), resulting from a false embryony. Robusta beans tend to be more rounded in shape. Especially large arabica beans, called margaropipes (or elephants) are also known. Peaberries and margaropipes are mainly of interest to speciality roasters.

These size grades are quantitatively defined from screen analysis in the laboratory. Sets of perforated plate screens are used, with the individual screens having apertures of specified dimensions: round holes for 'flat beans'; and, more infrequently, slotted holes for peaberries. Standardised screens are believed to have been devised first by the Jabez Burns Company in the USA for use in monitoring and assessing green coffee shipments from Central and South American countries; Gordon's of England have been and are a major supplier. Such screens with round holes are generally numbered from a so-called No. 21 down to No. 10 in whole unit steps (also sometimes half-units). These particular numbered sizes in fact represent hole sizes from a diameter of 21/64 to 10/64 in, at 1/64 in intervals (with 1/128 in intervals for the half sizes), which need to be very accurately drilled otherwise screening result differences will occur. In countries with entirely metric systems of measurement, numbering often corresponds to the diameter expressed in millimetres at values close to their inch equivalent and with similar intervals but with occasional small differences. However, the English numbering system is widely adopted.

Usually some four of these screens are used in a nest, in a given screen analysis determination, and are manually shaken, though sometimes a mechanical shaker will be used. Such a simple test will inevitably have variants of procedure and of screen dimensions at different locations. However, an International Standard[8] was published in 1980, harmonising existing practices and setting out nominal aperture diameters (English numbering, but specified in millimetres) and allowable tolerances for each numbered screen, which are, however, currently larger than desirable for consistency of results. This standard also sets out other necessary specifications for the screens to be used for green coffee beans; that is, overall size and surround depth, and spacing and pitch of the holes, in conformity with perforated screen specifications in other ISO standards for screens generally. The standard also details a recommended procedure for using the screens (manually). The sequence of aperture sizes is based upon so-called preferred numbers which correspond fairly closely in most cases with metric equivalents of 1/64 in units.

The results of a screen analysis of a representative commercial sample will generally fall as a good straight line (at least centrally) when aperture size (in millimetres) is plotted on a linear probability scale against percentage cumulative amount by weight held at each aperture size. The aperture size at which there is a 50% cumulative amount is an indication of the average size of the beans, and the slope of the line is an indication of the size distribution. Commercial specifications generally indicate the smallest aperture size through which all the beans will pass (together with a tolerance) and the smallest size at which, say, 99% of the beans are retained. A typical specification is that of Ivory Coast Grade 2, in which the coffee should all pass through a No. 16 screen (ISO, 6·30 mm diameter) with an allowable tolerance of 20% held, and all held by a No. 14 screen (ISO, 5·60 mm) with an allowable tolerance of 6% passing, and also all held by a No. 12 screen (ISO, 4·75 mm) with an allowable tolerance of only 1% passing.

A typical actual screen analysis for a sample of Ivory Coast Robusta grade II conforming to the specification is shown in Fig. 1, which also includes plots for grades I and III according to specification with tolerances. The lowest screen size used (No. 10 or 12) will pass broken bean fragments (French *brisures*), though this material may be assessed separately as a defect. It is sometimes recommended that all foreign matter should first be removed before screenings and the results of the screening analysis expressed for green beans only. There is a wide range

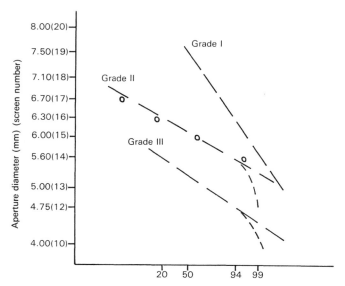

Fig. 1. Typical screen analyses for different size grades of Ivory Coast robusta green coffee beans according to specifications (including allowable tolerances), plotting percentage cumulative amount by weight (probability scale) held at each screen size (aperture diameter, mm) for screens according to ISO 4150. ○—○, Actual experimental data for grade II sample.

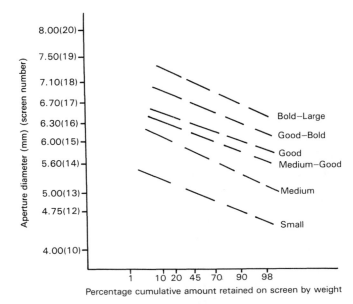

Fig. 2. Typical screen analyses for different size grades of Brazilian arabica green coffee beans according to specifications (including allowable tolerances). Percentage cumulative amount by weight held at each screen size (screens according to IS 4150) plotted on a probability scale.

of size grades for Brazilian coffees, a selection of which are shown in Fig. 2. Esteves[9] and Vincent[10] also report considerable data for screen analysis of green coffee.

2.4. Defects

'Defects' is a term used in commercial practice to describe the quality factor of the amounts of defective beans present and of extraneous matter (coffee and non-coffee) taken together or separately. This quality factor is more relevant, however, to dry processed than wet processed green coffees. The main types of defective bean are itemised in Table 2, with their French, Spanish and Brazilian equivalents.

In systems that involve physical counting, grades (types) are assessed according to the number of defects present per sample (300 g weight in metric countries; volume basis, 36 in^3 or approximately 1 lb US). However, not all defective beans or other defects are considered to make the same contribution. The New York Coffee and Sugar Exchange (NYCSE)

Table 2
Equivalent terminology for defective coffee beans and extraneous matter

English	French	Brazilian	Spanish
Dried coffee cherry	Cerise sèche	Grão em coco	Grano en coco
Black bean	Fève noire	Grão preto	Grano negro
Sour bean	Fève sûre	Grão ardido	Grano acre
Stinker bean	Fève puante	—	Grano fétido
Insect-damaged bean	Fève endommagée par insectes (piquée ou scolytée)	Grão brocado	Grano desperfecto par insectoes
Immature (quaker) bean	Fève immature	Grão verde	Grano verde
Broken bean	Fève brisée	Grão quebrado	Grano quebrado
Bean fragment	Brisure	—	—
Floater bean	—	Grão marinheiro	—
Shells	Coquilles	Conchas	Conchas
Husk	Coque	Casca	Cascara
Twig (stick)	Bois	Pau	Palo
Stone	Pierre	Pedra	Piedra

devised the 'black-bean-equivalent' count basis; that is, all other defects are assessed as requiring a number (from 1 to 15) of defective beans or other pieces of material present to equal one black bean. Brazil also uses this system. The French and OAMCAF systems are similar. The comparable ratings of some of the main recognisable types from some different national authorities are illustrated in Table 3; however, this information is continuously updated and primary sources should be consulted.

Extraneous matter, such as twigs (sticks or stalks from the coffee tree), stones, and even pieces of earth (or 'clods') at different sizes, is also separately counted, and the results are incorporated into the defect equivalency system, as illustrated in Table 4. Such matter is occasionally assessed in amount by weighing; in particular, foreign matter such as stones. Some sampling methods may not obtain representative samples for these materials, if, for example, they settle to the bottom of bags. In consuming countries, a more accurate picture for shipments will be obtained from the weight of residues in plant cleaning and destoning equipment.

The correct identifications of all these defects can be difficult, especially

Table 3
Equivalency ratings of green coffee bean defects in different countries[a]

Defective bean (English name)	USA (NYCSE) Brazils	USA (NYCSE) Centrals	Colombian	London Terminal Market Robusta	France, 1965[d]	Brazil (Institute of Brazilian Coffee)	Portugal, 1959	Indonesia (National Standard), 1982	Ethiopia (National Standard), 1973
Dried coffee cherry[b]	1	1	1	1	1	1	1	1	2
Black	1	1	1	1	1	1	1	1	2
Semi-black	—	1/2–1/5	1/2–1/5	1/2	—	—	—	1/2	1/2
Sour	1	1	1	1/2	1	1/2	1/2	—	—
Insect-damaged	1/5–1/10	—	—	1/2–1/5	1/10	1/2–1/5	1	1/5–1/10	1/2
Others									
Immature (quaker)	1/5	—	—	1/5	1/5	1/5	1/6	1/5	1/5
Floater (white light)	1/5	1/5	1/5	—	1/3	—	—	—	1/5
Bean in parchment	—	—	—	—	1/2	1/2	1/3	1/2	—
Broken (more than half)	1/5	1/5	1/5	1/5	—	1/5	1/6	1/5	1/10
Broken[c] (less than half)	1/5	1/5	1/5	1/5	1/5	1/5	1/6	1/5	1/5
Shells	1/3	1/3	1/5	1/5	1/5	1/3	1/6	—	—

[a] Figures in columns give defect (equivalent black bean) value for each one defective bean.
[b] Sometimes called 'pod'.
[c] *Brisure* in French.
[d] Generally includes OAMCAF countries.

Note. Other kinds of defective bean will also be rated in the systems of different countries, generally at low level. 'Stinkers' and 'foxy beans' rated at higher levels.

Table 4
Equivalency ratings of green coffee defects (extraneous matter—coffee/non-coffee)

Item (English name)	USA (NYCSE)			London Terminal Market Robusta	France, 1965[d]	Brazil (Institute of Brazilian Coffee)	Portugal, 1959	Indonesia (National Standard), 1982	Ethiopia (National Standard), 1973
	Brazils	Centrals	Colombian						
Husk fragments									
large	—	—	—	1/2	1	1	1	1	3
medium	1/2	1/2	1/3	—	—	1/2	1/2	1/2	1
small	—	—	—	1/5	—	1/3	1/3	1/5	1
Parchment fragments									
large	—	—	—	1/2	—	—	—	1/2	—
medium	1/2	1/2	1/3	—	—	—	—	1/3	1
small	—	—	—	—	1/3	—	—	1/10	—
Twigs (stalks or sticks)[a]									
large	2–3	2–3	2–3	5	2	5	3	5	10
medium	1	1	1	2	1	2	2	2	5
small	1/2–1/3	1/3	1/3	1/2	1/3	1	1	1	3
Stones,[b] pieces of earth									
large	2–3	2–3	2–3	5	—	5	3	5	10
medium	1	1	1	2	—	2	2	2	5
small	1/2–1/3	1/3	1/3	1/2	—	1	1	1	3

[a] Assessed by LTMR as 30, 20, 10 mm long, respectively.
[b] Assessed by LTMR as 10, 5, — mm diameter, respectively; not counted by France, but assessed by weighing.

from word descriptions only in different languages, as, for example, from the ISO vocabulary of coffee terms.[11] However, considerable work has been carried out in this field by Illy and colleagues.[12,13] Locally, reference sample beans can be used. Of interest also is the source through harvest, processing and storage of various defective beans, which is discussed in a number of texts.[7,14,15] The black bean, for example, mainly derives from a dead bean within the cherry on the tree; and the sour bean, together with the stinker bean, is associated with 'overfermentation' in the wet process. Of further interest is the effect of each of these defects on the flavour and other required qualities of green coffee, upon which the various defect equivalencies have been based and which no doubt will continue to be re-evaluated. The ISO Coffee Standards Sub-Committee is currently studying this subject with a view to harmonisation. The black bean is generally regarded as giving a 'heavy flavour', whereas the sour bean is regarded as especially important in downgrading flavour (sour taste together with 'oniony' flavour in Kenya) and even more so the 'stinker' bean (more likely to occur in wet-processed arabica than dry). Even insect-damaged beans can be assessed as producing some degree of flavour deterioration, though the physical appearance in the subsequent roast whole bean is the main negative factor.

Some of the main systems from different countries in which type numbers are allocated according to the quantity of defects are illustrated in Table 5, which applies almost entirely to dry processed coffees.

It will be noted that these long-standing systems differ somewhat in detail even among robustas of different countries. There is now a movement away from primitive counting methods over to weighing defects (total and individually) when adequate balances are available as this is simpler and more rapid. It is possible to provide approximate conversion factors to obtain allowable weights of defects from count specifications, provided the numbers of beans per 300 g or 1 lb sample is known (itself dependent upon the size grade concerned). For example, five actual blacks in an Ivory Coast grade 2 Coffee at, say, 2100 beans per sample equates to 0·24% by weight approximately; whereas 15 defects as insect-damaged (150 actual) in any sample of 3000 beans would equate to 5% by weight approximately. The type specification numbers have been devised by the various marketing organisations themselves; though a number of countries (e.g. Kenya, India, Indonesia) have also set out their individual country specifications within national standards. It should be pointed out that all these types are not all necessarily available in a given crop year, and that production of one or more types will generally predominate. Qualities

Table 5
Type numbering of green coffee in various countries by maximum defects

Country	Type no./ description	Maximum allowable number of defects per sample based on:	
		300 g	1 lb
Brazil[a]	NY 1/2, Fancy	4	—
	NY 2/3, Extra	8	9
	NY 3/4, Prime	19	21
	NY 4, Base	26	—
	NY 4/5, Superior	36	45
	NY 5/6, Good	79	90
Ivory Coast[b]	Extra Prima	15	—
	Prima	30	—
	Supérieure (Base)	60	—
	Courante	90	—
	Limite	120	—
UK Terminal Market Robusta[c]	Base	—	150
	Untenderable	—	>350
Indonesia	1	11	—
	2	12–25	—
	3	26–44	—
	4	45–80	—
	5	81–150	—
	6	151–225	—
Tanzania Robusta[d]	FAQ	10% by count; 5% by weight	
	UGQ	>10% by weight	

[a] The official Brazilian classification system includes a much more detailed breakdown into types, e.g. 30 defects = NY 4–10, and goes down to NY 8.
[b] Also includes restriction as number of *brisures*, i.e. maximum of 5, 5, 10, 15, 15 assessed as defects, and also of blacks together with maximum allowance of stones (1·25 g, i.e. 0·4% by weight).
[c] Also includes a wider range above and below base.
[d] Also includes restrictions as to numbers of blacks (3% by count), dried coffee cherries (1%) and amount of extraneous matter (0·5% by weight).

above and below 'base' are usually subject to a small bonus or discount payment in selling/tendering.

With wet processed coffees, the number of defects is generally quite low compared with those of Table 5; for example, the NYCSE system provides for a maximum of not more than 13 defects per 1 lb sample from Central/Latin American coffees corresponding to the name 'Extra', so that inferior types are not marketed. With both kinds of coffee, reduction in the number of defects is progressively continuing, with efficient hand-picking techniques and the advent of electronic sorting machinery for removing blacks, other discoloured beans and even 'sours' (by fluorescence detection).

The test methods, involving relatively simple procedures, have grown up within the individual countries. The International Standards Organization has harmonised many of these methods; see, for example, reference 16 on the olfactory and visual examination and determination of foreign matter and defects, and reference 17 on the determination of insect-damaged beans.

2.5. Colour

The overall colour of green coffee beans from different sources clearly can differ considerably and is a further indicator of quality. Wet processed arabicas are generally blue–green to grey–green and with uniform colour, with the blue–green colour especially prized; whereas dry processed robustas tend to be brownish and non-uniform. Storage of green coffees will cause changes in colour appearance; thus dry processed Brazils become whitish on prolonged storage.

Green bean colour and appearance in Kenya coffees are contributing factors in their system for establishing standards as already described. Brazil has a classification scheme for colour, in the range green–greenish–medium–lightish–light–yellowish–yellow. The ISO[16] recommends the classification terms bluish–greenish–whitish–yellowish–brownish.

2.6. Roasting Characteristics

The appearance of the green coffees after sample roasting (generally to a light or medium colour) is also used for the assessment of quality. Again, Kenya incorporates an assessment in its quality standard scheme. Brazil provides also an assessment in terms of the number of defects observed per 100 g of the roasted beans, with the characterising terms 'fine roast' (0–1 defects), 'good to fine' (2–5), 'good roast' (8–10), and 'poor' (more than 12). The defects will particularly include so-called pale beans, but also those markedly dissimilar from the average in colour, and insect-damaged beans.

2.7. Bulk Density
The bulk density of green coffee is a readily measurable quantity (either by free-flow or vibrational methods). This figure is of significance in relation to the bulk density of roasted coffee and therefore to volume in packages.

2.8. Crop Year
The crop year is a further factor in flavour and other quality assessments. Some differences can be expected in fresh crops from one year to another in different countries, and in main crop and subsidiary crops within one year. Differences will also be found in a 'new crop' and 'old crop' (resulting from storage); for example, in the presence of 'faded beans' with a 'thinner' and often 'woody' flavour.

3. STORAGE

3.1. Storage Conditions
Green coffee, following its preparation by the methods described in Chapter 1, will inevitably be stored under different conditions for different periods of time at different locations, and indeed its transportation from the producing country to the consuming countries involves a period of enforced storage, which may be many weeks. During storage some changes in the quality and appearance of the green coffee can be expected (including the potential development of moulds); though these changes will be strongly dependent upon the ambient temperature and relative humidity, the latter influencing the moisture content of the green coffee, and of course the time of storage. Green coffee after drying can be stored in bags at the point of final processing and grading, where the temperature/humidity conditions may well be favourable; and also in warehouses at the point of embarcation (which, being low lying, may not be so favourable, with extremes of temperature and humidity). Ship transportation to the ports of consuming countries may involve the problem of moisture release (as in containerised shipments) as a consequence of a change to a lower humidity rather than temperature (see section 3.2 on isotherms), and this moisture may deposit locally and cause mould growth. There may well be storage at the port of the consuming country, before road or rail transport to the point of roasting; or indeed

Table 6
Published data showing values of %ERH for various green coffees at 12% moisture content (w/w) at different temperatures

Green coffee type	Isotherm mode	Temp. (°C)	%ERH	Method of moisture determination	Reference
General	—	25	60	—	Sivetz and Desrosier[7]
Kenya	Ads./Des.	28	63	Kenya[18]	Stirling[20]
Brazils	—	28	77	—	Quast and Texeiro[21]
Arabica	Ads.	25	68		
	Des.	25	71	Air oven	
Arabica	Ads.	35	69	(105 °C	Ayerst[22]
	Des.	35	72	for 4 h)	
Robusta	Ads.	25	67		
	Ads.	35	69		
Arabica	Sorp.	30	63	Guilbot[19]	Multon[23]

there may be extended storage before sale in the spot market (e.g. at the Brazilian coffee stores at Trieste). Storage may well span the period between the beginning of a new crop and that of the next. Wet processed green coffee may also be stored, not in the form of beans but as parchment coffee, especially where long periods of storage are expected. In the parchment form, green coffee is more stable to unfavourable temperature and humidity conditions.

3.2. Isotherms

Isotherms show the relationship between moisture content and equilibrium relative humidity for a given temperature. Isotherms for green coffee have been fully discussed in Volume 1, Chapter 2. Table 6 reproduces some of the experimental data available, in particular the percentage equilibrium relative humidity corresponding to a 12% w/w (or as is) moisture content in green coffee, generally regarded as close to the acceptable limit for stored and shipped green coffee. Unfortunately, methods of moisture determination differ; in Table 6, those methods by Stirling and Multon are to be preferred, and by implication the data.

3.3. Storage Stability

Considerable studies have been made of the storage stability of green coffee in different temperatures/environments, e.g. by Stirling[20] and by

Multon,[24] and the data compiled and reviewed by Clarke.[25] Potential spoilage by moulds is a particularly important commercial consideration. It is generally considered that mould growth on green coffee is likely to start with moisture contents in excess of 13% w/w. Such a level corresponds to a percentage equilibrium relative humidity of 67% (Stirling's data[20]) at 25°C, which is consistent with other general information that a water activity of at least 0·65 is needed to initiate mould growth and optimally at 0·85. Deterioration in quality is generally associated with the presence or otherwise of free water, though this figure is difficult to ascertain for green coffee. Green coffee is in fact in a mild state of respiration during storage, with the evolution of some heat.

Both Stirling and Multon found that wet processed arabica coffee should keep close to its original quality for six months provided that on storage the coffee had first been dried (in its parchment form) to a moisture content of not more than 11·0% w/w (Stirling) or 13% dry basis (= 11·5% w/w, Multon), that the relative humidities of the air are in the range 50–70% (Stirling) or 50–63% (Multon), and that the air temperature is maintained at 20°C or less (Stirling) or 26°C or less (Multon). Additionally, Stirling found that controlled ventilation was more beneficial than sealed conditions, especially at the higher temperatures. He also found that hulled but not cleaned (i.e. with retention of silverskin) coffee showed greater stability, and in particular parchment coffee. It is interesting that Multon indicated that some length of storage was favourable to quality, which may well be related to the formation of suitable aroma precursors. Some storage conditions provide flavour qualities which are desired among some consumers, e.g. 'aged' coffees from Venezuela, and 'monsooned' coffees in India. There are also trade distinctions of 'new crop' coffee (freshly harvested and processed), 'second crops', and 'old crops' (necessarily stored for a longer period in the producing country), which find separate acceptance.

There are therefore complications in precisely defining a shelf-life for green coffee, though the criteria of Stirling and Multon provide a useful guide. In practice, the conditions of temperature and humidity are likely to vary widely over the actual period of storage. Flavour quality changes would need to be integrated over the whole temperature/humidity regime.

3.4. Methods of Storage
Green coffee is transported and temporarily stored in bags, usually made of jute (though other natural fibres have been used, and also polypropylene). The net contents are generally 60 kg, i.e. from Brazil, Kenya and African

producers of robusta coffee; but 70 kg is common to Colombia and Central/Latin American countries generally. A number of other countries, however, bag coffee at even higher weights (75, 80 and even 90 kg). For the purpose of ICO statistics on green coffee, exports/imports of green coffee are recorded in 60 kg units. Bag sizes were originally determined by what a man could reasonably carry on his back.

Movement and stacking of bags are conveniently carried out by use of pallets. Sivetz and Desrosier[7] describe the typical US practice of pallets measuring 5 ft × 6 ft × $\frac{1}{2}$ ft accommodating 20 bags in four layers piled in alternate directions. Stacks are usually then up to four to five pallet loads high, i.e. some 100 bags. Jute bags are preferred to polypropylene for pallet storage for reasons of non-slippage.[6]

Silos (of steel or galvanised iron) are also conveniently used for short-term storage within a large roastery; that is, immediately before roasting. Such silos are often multi-compartmented with a cone bottom section, for discharge of different coffee types as required for batching. Large-sized silos for bulk storage in the consuming country are less common; though more so in producing countries before bagging off. Some of the factors that contribute to their good design are discussed by Stirling; for example, ventilating systems and the use of insulated structures, especially to diminish the effect of the direct rays of the sun on the roof. There may be problems of segregation of large from small beans.

Air conditioning, including injection of cold air from refrigerating machinery, has been discussed by a number of authors, e.g. Sivetz and Desrosier[7] and Bauder.[26] Sivetz and Desrosier offer a figure of 40 tons refrigeration for a warehouse holding some 8000 bags of green coffee.

4. PRE-TREATMENTS

4.1. Cleaning and Destoning

Before blending and roasting, it is often necessary to remove the small fractional percentage quantity by weight of extraneous matter that may be present in the green coffee delivered in bags (already discussed on page 43). This extraneous matter, some of which may have settled to the bottom of the bags, may comprise a whole range of heavy foreign materials such as stones, nails, coins, light contaminants, wood splinters, fibrous materials, and general dust. As also mentioned, silos are generally used for short-term storage, and it is convenient to tip the contents of the bags (of the same kind of green coffee in one sequence) through a grill (hooded) on to conveyors leading to separate or combined cleaning (usually by air currents to cyclones) and destoning equipment. Sivetz and Desrosier[7] provide detailed descriptions of such equipment, including the Gump cleaner (though apparently no longer sold), which removes both light and heavy material. In large installations, the cleaned green beans are air-veyed to their appropriate silos. It is convenient to be able to weigh the total extraneous material collected as a percentage in a consignment of green coffee, since this figure will more truly reflect the amount of this material, rather than that obtained by bag sampling techniques. Since roasted coffee beans are considerably less dense than green beans and therefore stones, destoning is often more conveniently carried out or completed subsequent to roasting.

4.2. 'Health' Coffees

In Germany particularly, the 'indigestibility' of normal coffee brews has been an issue of continuing interest, reflected in so-called commercially manufactured 'health' or *Sänft* (= smooth) coffees, both roast and ground and instant, in which components reported or alleged to be responsible for gastrointestinal irritation and other symptoms in some consumers (including emetic activity) are removed from the green coffee.[27]

Two main processes have been used, first, the dewaxing of green coffee, and secondly the old Lendrich patented steaming process.[28] Dewaxing, that is, of the outer layers of the green coffee, will occur simultaneously with solvent decaffeination of green coffee, as described in the next chapter. Less rigorous washing by solvents, however, is used to remove only the waxy substances (see Volume 1, Chapter 5), of which the carboxy-5-hydroxytryptamides (C-5-HT) are a component and are used as an analytical parameter of efficiency for the removal of undesired components. Van der Stegen has determined the quantity of C-5-HT present in untreated and in the same dewaxed coffee, in the original green, roasted coffee and in a household brew.[27] He found that in untreated coffee, 22% was decomposed on roasting; 70% appeared in the spent grounds of the brew, with some 6% in the brew, though all would be retained by a paper filter. A dewaxing process removed 64% of the initial C-5-HT and the roasting a further 12%; on brewing, 26% was present in the spent grounds, and only 1% in the brew. Van der Stegen reports that the C-5-HT is of main interest in relation to its roast products (which have been examined by Viani[29]), some of which are phenolic and water soluble. Removal of the waxy layer before roasting will thus result in a reduction of such phenolic products in the brew, as an explanation of the beneficial results claimed.

The Lendrich process is a low-temperature steaming operation on the green beans. Windemann, as reported by Van der Stegen,[27] found 3-methoxy-4-hydroxystyrene in the steam condensate, which is formed by the decomposition of ferruloylquinic acid in the chlorogenic acid complex of green coffee (see Volume 1, Chapter 5).

Further information on the compounds in the condensate were given by Van der Stegen.[27] Changes in the chlorogenic acid complex (determined by GC methodology) on steam-treating green coffees were compared with the untreated samples, and also for the two roasted coffees in respect of breakdown products especially phenols; a general reduction in quantity was found.[30] Changes in the chlorogenic acids on roasted steam-treated (2 h at atmospheric pressure) coffee were also reported by Van der Stegen and Van Duijn.[31]

5. SELECTION AND BLENDING

5.1. Availability

Of the marketed green coffees described in section 2 there is some seasonal availability. A crop year is deemed to commence in July and to finish in the July of the following year. Within the crop year, harvesting of the main crop will generally take place at specified periods lasting a few months, often followed by subsidiary crops. Green processing may take a few months, so that with shipment time a harvested crop may take several months to reach Europe or the USA. Table 7 gives average harvesting dates for a number of coffees.

5.2. Selection

The green coffees imported from the various producing countries as described in section 2 (see also Volume 1, Chapter 1) are the basic raw materials for the manufacture of roasted and of soluble coffee. Substantial manufacture, especially of soluble coffee in Brazil, takes place in the producing countries themselves, and therefore mainly uses the indigenous green coffee.

The selection of green coffee from those available for roast (and ground) coffee is primarily a matter of assessing required contributions in a blend for consumer taste requirements at acceptable cost. However, a number of roasters also market individual coffees, designated by country of origin (e.g. 100% Colombian coffee). There are various principles of blending that can be enunciated, typically described by Davids;[32] together with careful consideration of cost factors in the final product, which include

Table 7
Harvest and exporting times for green coffees for various countries

Country/district	Type	Harvesting	Export from origin
Colombia			
Caldas, N.	Manizales, Armenian	Oct.–Dec.; April–June (s)[a]	Manizales, Armenian,
Caldas, S.	As above	March–May; Oct.–Dec. (s)	Medellin (MAMS)
Antiquoia	Medellin	Oct.–Jan.; March–May (s)	Excelso all year round
Cauca	Popayan	April–June; Dec.–Feb. (s)	
Mexico	Prime washed	Nov.–Feb.; Sept.–Oct. (s)	Sept.–Jan.
Costa Rica			
Atlantic Zone	Low-grown and high-grown Atlantic (LGA and HGA)	Aug.–Dec.	Sept.–Dec.
Central Plateau	Hard, good hard and strictly hard bean (HB, GHB, SHB)	Nov.–Feb.	—
Kenya			
Kiambu, Thika and Ruiru	All qualities	Oct.–March	All year round according to quality
East of Rift		June–Jan.	
West of Rift		Aug.–Feb.	
Brazil			
Santos	All types	June–Sept.	
Parana			
Minas		April–Aug.	
Bahia			
Ivory Coast	Robusta grades 1–3	Oct.–April; July–Nov. (s)	All year round

[a] s = Second crop.

roasting. Various combinations of 'milds' may be used, but Brazils (soft qualities) are generally included as they are characterised by their blendability and lower cost (e.g. 50:50 blends). Smaller roasters may apply special considerations as regards the selection of margaropipes and other large-sized beans, or of peaberries only and mocha coffees, which are in small supply. National and regional tastes are important in selection, so that, for example, the cheaper green robusta coffees, usually more darkly roasted, are very popular in France and Italy. Swedish, Finnish and German manufacturers probably offer the highest proportion of high-quality coffees for their markets, despite the fact that in Germany there have been and are considerable extra governmental taxes on coffee.

While manufacturers attempt to maintain marketed brands of roasted (and ground) coffee, as with tea, at a consistent flavour quality, blending offers flexibility of actual formulation. This is necessary in view of changes of availability of particular coffees during the year, changes of flavour quality by crop or storage, and relative price changes. Formulations are generally secret, though arabica contents are often declared (e.g. in France they are legally required). Some indications of average blends used may be obtained from import statistics of green coffee for different countries over different years, as in Table 8. Such information does not, however, so readily indicate average roast coffee blends, where distortion due to a high proportion of coffee for instant coffee exists.

Similar principles of blending apply, however, in the manufacture of instant coffee, whether powder, granules or freeze-dried, again with the

Table 8
Approximate percentage arabica/robusta green coffee imports of various countries, 1981. (Source: ICO statistics.)

Country	Total % Arabica[a]	% Brazils only
UK	35	14
Germany	93	9
France	42	18
Italy	60	35
Spain	75	21
Sweden	95	25
Switzerland	78	23
World	75	30

[a] % Robusta obtainable by difference of total percentage arabica from 100.

important consideration of cost factors. Quality factors are very much determined by the overall process used, and these have become increasingly sophisticated especially in retaining flavour volatiles. The older processes of simple spray-drying, or even drum-drying, would not always justify the use of the better qualities rather than the poorer. Robusta coffees are widely used in blends of instant coffee, dependent upon market brand factors of cost and consumer taste requirement in different countries. The proportion of soluble to roast coffee manufacture differs widely in different countries; so that whereas in the UK and Japan the proportion (from consumer sales data) is of the order of 90%, in other European countries it is much lower (i.e. a few per cent in Scandinavian countries) and in the USA of the order of 25% (though the quantities consumed are large).

5.3. Blending Methods

No special arrangements are normally made for blending, since the roaster itself provides the necessary blending action for the different green coffees, assuming that they have been correctly weighed in and are therefore in the required proportions. Sometimes the individual green coffees are first separately roasted, and then it is necessary to have separate blending equipment before grinding and packing, or before extraction in instant coffee manufacture.

REFERENCES

1. Jobin, P., *The Coffees Produced throughout the World*, 4th edn, Jobin, Le Havre, 1982.
2. Rothfos, B., *Coffee Production*, Gordian–Max Rieck, Hamburg, 1980.
3. Clarke, R. J., *Fd Chem.*, 1979, **4**, 81–96.
4. International Standards Organisation, ISO 4072–1982; also BS 6379, Part 1, 1983.
5. International Standards Organisation, ISO 6666–1983; also BS 6379, Part 3, 1983.
6. Marshall, C. F., in *Coffee: Botany, Biochemistry and Production of Beans and Beverage*, Eds M. N. Clifford and K. C. Willson, Croom Helm, London, 1985.
7. Sivetz, M. and Desrosier, N. W., *Coffee Technology*, AVI, Westport, Conn., 1979.
8. International Standards Organization, ISO 4150–1980; also BS 5752, Part 5, 1981.
9. Esteves, A. B., *J. Coffee Res.*, 1972, **2**(3), 4–20.
10. Ballion, P., Hahn, D. and Vincent, J-C., *Café Cacao Thé*, 1973, **XVII**(3), 231–40.

11. International Standards Organization, ISO 3509–1976, with amendments DAM 3509; also BS 5476, 1977.
12. Illy, E. and Ruzzier, L., *Proc. 3rd Coll. ASIC*, 1967, 24–45.
13. Illy, E., Brumen, G., Mastropasqua, L. and Maugham, W., *Proc. 10th Coll. ASIC*, 1982, 99–128.
14. Sivetz, M., *World Coffee and Tea*, 1971, 15–28.
15. Haarer, A. E., *Modern Coffee Production*, Leonard Hill, London, 1956.
16. International Standards Organization, ISO 4149–1980; also BS 5752, Part 4, 1980.
17. International Standards Organization, ISO 6667–1985; also BS 5752, Part 7, 1986.
18. Wootton, A. E., *Proc. 3rd Coll. ASIC.*, 1967, 92–100.
19. Guilbot, A., *Proc. 1st Coll. ASIC.*, 1963, 24–32; also *Café Cacao Thé*, 1963, **7**, 49–56, 192–200.
20. Stirling, H., *Proc. 9th Coll. ASIC.*, 1980, 189–200.
21. Quast, D. and Texeiro, N., *Fd Technol.*, 1976, **30**, 598–600.
22. Ayerst, G., *J. Sci. Fd Agric.*, 1965, **16**, 71–8.
23. Multon, J. L., *Proc. 8th Coll. ASIC*, 1974, 65–72.
24. Multon, J. L., *Proc. 6th Coll. ASIC*, 1973, 268–77.
25. Clarke, R. J., in *Handbook of Food and Beverage Stability*, Ed. G. Charalambous, Academic Press, New York, 1986.
26. Bauder, H. J., *Proc. 6th Coll. ASIC*, 1974, 278–83.
27. Van der Stegen, G. H. D., *Fd Chem.*, 1979, **4**(1), 23–30.
28. Lendrich, P., German Patent DR 576, 515, 1933.
29. Viani, R., *Proc. 7th Coll. ASIC*, 1975, 273–8.
30. König, W. A. and Sturm, R., *Proc. 10th Coll. ASIC*, 1982, 271–8.
31. Van der Stegen, G. H. D. and Van Duijn, J., *Proc. 9th Coll. ASIC*, 1980, 107–12.
32. Davids, K., *The Coffee Book*, Whittet Books, Weybridge, pp. 21–59.

Chapter 3

Decaffeination of Coffee

S. N. KATZ
*General Foods Manufacturing Corporation,
Hoboken, New Jersey, USA*

1. INTRODUCTION

Many of the physiological effects of coffee beverages are due to their caffeine content. The two major species of coffee, arabica and robusta, contain approximately 1 and 2% caffeine respectively. This corresponds to a cup of regular coffee containing on the average 85 mg of caffeine with a range of 50 to 150 mg based on preparation, blend and cup size.[1] A cup of instant coffee contains approximately half that of regular coffee with a reported range of 40 to 110 mg. To minimise the physiological effects and still obtain the desirable attributes of a coffee beverage, many procedures for caffeine removal have been proposed.

The process of removing caffeine from coffee was invented[2] by Dr Ludwig Roselius and Dr Karl Wimmer of the Kaffee-Handels-Aktien-Gesellschaft in Bremen, Germany. This invention is described in the German patent letter, DRP Deutsches Reich Patentschrift, dated from 1905. A commercial decaffeination process was implemented by the above company and they marketed a 'caffeine-free' coffee under the name of Kaffee Hag.

Around 1912, Roselius extended his business to the United States, and a coffee extraction plant was built at New Brunswick, New Jersey. During the First World War the American Kaffee Hag business was expropriated by the American government and the ownership of the company passed over to one Mr Gund, who moved the factory to Cleveland, Ohio. Mr Gund later sold his business to the Kellogg Company and the activities of the company were transferred to a 'million dollar plant' at Battle Creek, Michigan.

After the armistice, Roselius returned to the United States and started a new company, the Sanka Coffee Corporation. At first this company imported decaffeinated green coffee from Germany and roasted it in New York. Around 1927 Roselius made a deal with General Foods Corporation (The Postum Company) to conduct the Sanka business on a partnership basis. General Foods financed and equipped the Sanka factory in Brooklyn, New York. In 1932 Roselius sold his interest in the Sanka Coffee Corporation to General Foods Corporation, and General Foods became sole owner.

In December of 1937, General Foods Corporation purchased the American Kaffee Hag Company from the Kellogg Company, thereby gaining complete control of decaffeinated coffee in the United States and Canada at that time. About 1955 the Nestlé Corporation and others introduced decaffeinated coffee in the United States. In the 1970s General Foods Corporation bought the Kaffee Hag Company in Germany from the heirs of the aforementioned Dr Roselius.

Commercial decaffeination of coffee has mostly been performed on green coffee beans prior to roasting in order to minimise flavour and aroma problems. The desirable organoleptic properties of the beverage

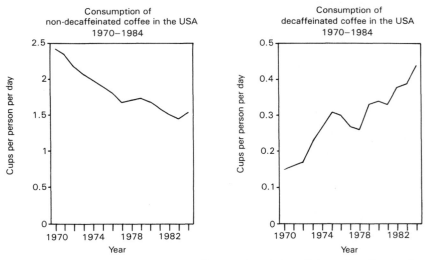

Fig. 1. Consumption of non-decaffeinated and decaffeinated coffee in the USA.

are developed after roasting. Allowing the decaffeination process to be performed on the green bean minimises aroma and flavour changes and losses. In addition, the two most common techniques are either based on solvent decaffeination or water decaffeination. Patent literature, however, also cites caffeine removal from roasted coffee extract.[3,4]

With the increased emphasis on health and health products and the known physiological effects of caffeine in coffee, more people are switching to beverages with less or no caffeine. As shown in Fig. 1 depicting the *per capita* consumption of coffee in the USA (approximately 50% of all coffee consumed), the number of cups of undecaffeinated coffee has been on a steady decline since 1970.[5] However, the *per capita* consumption of decaffeinated coffee has increased 153% in 12 years. The ratio of undecaffeinated to decaffeinated cups per person has dropped in this period from 16:1 to 4:1. With this continuous strong trend, the drive for the production of high-quality decaffeinated coffees produced in an economical and safe environment is stronger than ever.

2. SOLVENT DECAFFEINATION

Decaffeination of green coffee beans is still the preferred methodology used by most commercial coffee companies. The most direct approach to accomplish this is to contact pre-wetted green coffee with an appropriate solvent in conventional equipment in either a batch or continuous mode. Many solvents have been suggested for use in decaffeination of coffee beans or water extracts. Some of the criteria for the choice of solvent include the following:

(1) safety,
(2) cost of solvent,
(3) caffeine solubility,
(4) caffeine specificity,
(5) ease of solvent removal and recovery,
(6) toxicity and chemical reactivity,
(7) environmental effects.

Originally, benzene was used as the extraction solvent, but when trichloroethylene became available at reasonable prices, benzene was replaced by this non-flammable and less toxic solvent. As other chlorinated solvents were made available on a commercial basis, many coffee manufacturers switched to methylene chloride (dichloromethane), which has a higher caffeine solubility and a lower boiling point, allowing easier removal from the beans. Trichloroethylene has been under investigation by the US Food and Drugs Administration since 1976, and the only chlorinated solvent used either directly or indirectly is methylene chloride.

Table 1 lists most of the solvents that have been mentioned in US patents. Other than methylene chloride, the only known commercial solvents used are ethyl acetate,[6] coffee oil or other triglycerides,[7] and supercritical carbon dioxide.[8] The last of these will be discussed separately due to its unique processing requirements.

The basic principles of solvent removal of caffeine from green coffee are similar for any of the commercial solvents. The processes differ mainly in the methods of caffeine removal from the solvent and the methods of regenerating and recycling the solvent. If a volatile solvent is used, simple evaporation, usually in multi-effect or other low utility requirement evaporators, results in a clean solvent for recycling, leaving a caffeine-rich sludge to be refined. If a high-boiling solvent is used, either a water wash in a countercurrent liquid–liquid extractor or a high-vacuum sublimator is required, both to regenerate the relatively expensive solvent and to recover the valuable caffeine. Unlike most solvent extraction or degreasing operations the decaffeination of coffee beans requires the addition of significant quantities of water. A minimum water content of approximately 20% is required to be able to solvent-decaffeinate, and as much as 55% or saturation is used. Below 20% only limited decaffeination can occur no matter how much solvent is used.[9] Explanations for this

Table 1
Selection of solvents mentioned in United States patents

Acetic ether	Ethylene chloride
Alcohols	Fatty acids
Aliphatic hydrocarbons	Isopropylchloride
Benzene	Ketones
Carbon dioxide	Methylene chloride
(liquid and supercritical)	Mineral acids
Carbon tetrachloride	Mixed halogenated hydrocarbons
Chloroform	Paraffin oils
Coffee oil	Sulphur hexafluoride
Dichlorobenzene	Tetrachloroethylene
Dichlorobenzol	Toluene
Dichloroethylene	Trichloroethylene
Dimethylsulphoxide	Various fluorinated hydrocarbons
Esters (e.g. ethyl acetate)	Vegetable oils

phenomenon include softening and opening of the cellular structure to allow the caffeine to diffuse out of the bean. A more likely explanation involves the ability of caffeine to bond to the plentiful chlorogenic acid and its potassium salt, potassium chlorogenate, a complex ion-salt-like material that is essentially insoluble in the non-polar solvents commonly used.[10] Exceptions to this rule are aprotic solvents, but other problems preclude their use.[11]

A typical solvent extraction process is shown in Fig. 2. In this semi-continuous process, fresh green coffee beans are steamed for half an hour, resulting in a moisture content of 16 to 18% w/w in the first column of a battery of columns. This is followed by a pre-wetting step to increase the coffee bean moisture to above 40% by weight. The wetted coffee beans are then countercurrently extracted by a solvent at temperatures between 50 and 120°C (120–250°F). The column from which most of the caffeine has been removed (95–98% decaffeinated) is isolated, solvent-drained and then steam-stripped to remove all residual solvent. Decaffein-ated green beans are then discharged from the column and immediately dried before storage. Caffeine-rich solvent exiting from the fresh side of

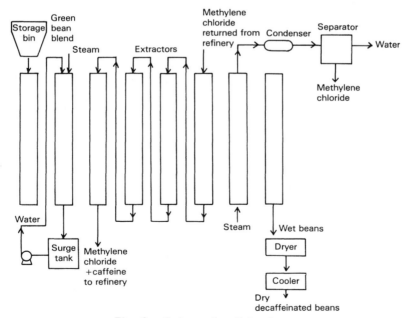

Fig. 2. Solvent decaffeination.

the column battery is then treated as discussed previously for caffeine recovery and solvent regeneration. Care must be taken to ensure that the solvent is saturated with water prior to recycling, in order not to dehydrate the pre-wetted coffee beans.

Typical operating conditions as stated in US Patent 3,671,263 are as follows:[12]

Steaming	30 min at 230°F
Pre-wetting	150°F to 42% water by weight
Caffeine extraction	Methylene chloride at 150°F for 10 h; 4 lb methylene chloride per pound of green coffee beans
Steam stripping	1·5 h

3. WATER DECAFFEINATION

A different approach to decaffeination of coffee beans was invented in 1941 by Berry and Walters of the General Foods Corporation.[13] This invention is based on using water to remove the caffeine from the bean. To prevent the extraction of all the water-soluble components of coffee beans, in excess of 20% by weight, the extraction water contains essentially equilibrium quantities of the non-caffeine soluble solids but no or very little caffeine. Claimed advantages include higher extraction rates, elimination of water-insoluble waxes extracted by the solvent, purer caffeine in a caffeine recovery system, and less heat treatment of coffee beans by the elimination of the solvent-stripping step since there is no contact between solvent and beans. An additional simplification in processing is the removal of the pre-steaming and pre-wetting steps, because all the water required to free the caffeine within the bean is supplied by the water extract. A flow sheet of the water decaffeination process is shown in Fig. 3. As in many coffee processing operations, use is made of a battery of columns in which green coffee beans are countercurrently extracted by a water extract containing about 15% solids other than caffeine. Water is absorbed preferentially by the dry beans during their residence in the extractors, and the solids concentration may increase to as high as 30%. Bean residence of approximately 8 h is required to remove about 98% of the original caffeine. Decaffeinated wet beans have about 53% by weight of moisture, and adhering solubles are washed with fresh water to minimise drying problems as well as to recover

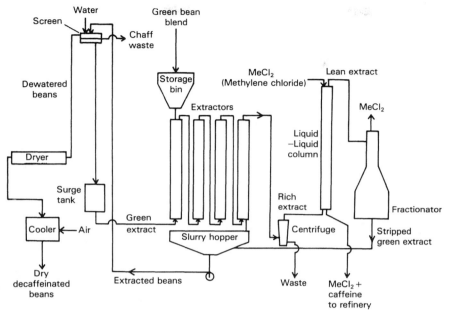

Fig. 3. Water decaffeination.

solids. This wash water is added to the caffeine-free water extract to bring the solids concentration back to the starting level of about 15%. Finally, the washed decaffeinated green beans are air-dried to be sold as a source of decaffeinated beans or processed in a conventional manner to produce decaffeinated regular roasted coffee or decaffeinated instant coffee.

The extract, rich in caffeine (about 0·5% by weight), must be recycled after caffeine is removed in order to maintain the solids balance with other soluble solids contained in the green coffee beans. Caffeine is usually removed preferentially by contact with a specific solvent, but some newer patents have indicated improved developments in the use of solid adsorbents that have been suggested for at least 50 years. Caffeine-rich extract, after centrifuging to remove suspended solids such as coffee chaff, is contacted with an immiscible solvent such as methylene chloride to remove caffeine selectively. Any other solids extracted by the solvent are usually lost in the caffeine refining system. Various liquid–liquid contactors (extractors) have been tried, the most effective being a rotary disc contactor[14] to reduce the caffeine content of the green coffee extract to below 0·05% caffeine by weight. Countercurrent contact is performed by allowing the caffeine-rich extract to rise through the dense methylene chloride continuous phase at temperatures around 80°C (180°F) to take advantage of higher caffeine distribution coefficients and reduced emulsification problems at elevated temperatures. Extraction efficiency is controlled by the effect of the rotary disc speed on droplet size up to the point of column flooding. A solvent-to-extract ratio of about 4 to 8 is required to remove 99%+ of the caffeine from the extract. After solvent contact, the lean green coffee extract has dispersed and dissolved solvent in it that must be removed before it is allowed to contact coffee beans. Solvent removal is performed in a distillation column with feed to the top tray and reboiling and direct steam injection at the bottom to ensure complete solvent removal. Overs from the distillation column are at the azeotropic composition of the solvent, 98·5% by weight for methylene chloride, unless excess steam is used. Solvent, rich in caffeine exiting from the liquid–liquid column, is recovered by evaporation. Caffeine refining is discussed elsewhere. It is extremely important that all waste streams be very low in solvent content to meet environmental and legal requirements.

The use of adsorbent to remove caffeine in the past has not been economically useful as it has altered the quality of the final coffee detrimentally. A more recent claim overcoming these problems has been published in the European Patent Office by Coffex AG. This process utilises activated carbon that is preloaded with other coffee extract substances or with substitute substances of similar molecular structure or size, especially with carbohydrates, e.g. cane sugar, in order that the charcoal will take up as few extracted substances as possible. It is claimed for this procedure that, apart from caffeine, practically no other coffee components are removed from the green coffee. Decaffeination of the beans can be carried out either by total soluble solids extraction, decaffeination and reinfusion into the bean or by the use of the green extract procedure described above for a solvent system.

4. SUPERCRITICAL CARBON DIOXIDE DECAFFEINATION

The use of the solvent supercritical carbon dioxide for decaffeination of coffee, commercialised by HAG–GF in Germany, is based on the 1970 patent of Studiengesellschaft Kohle of Mulheim, Germany. Its use for decaffeination of green coffee beans is very similar to that of other solvents, with the major advantage of the use of an inert gas compressed

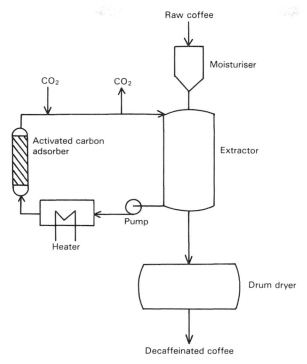

Fig. 4. Carbon dioxide decaffeination process, schematic.

to pressures high enough to increase caffeine solubility sufficiently for use in a closed cycle, as shown in Fig. 4. The major disadvantage is also due to the high pressures requiring costly equipment and batch processing. Bean preparation is basically the same as for other solvents, i.e. mechanical cleaning to remove dust and chaff followed by steaming and wetting to increase the moisture to a 30–50% range. The decaffeination process consists of loading the pre-moistened beans into the extractor vessel and at the same time solid adsorbent is loaded into the adsorption vessel. Sufficient carbon dioxide is loaded into the vessels and the circulation pump is started. The heat exchanger is used to increase temperature, thereby also increasing the pressure to the desired extraction level. As the carbon dioxide flows through the vessels, caffeine is extracted from the moist beans, and caffeine-rich carbon dioxide is recycled to the adsorber where caffeine is adsorbed on to the activated carbon. The caffeine-free carbon dioxide then starts a new cycle via the heat exchanger to the extractor. Processing conditions are selected to minimise the extraction time to increase plant capacity and to improve process economics. The average concentration of caffeine in the carbon dioxide is low, requiring large quantities of carbon dioxide to be circulated. Therefore for energy conservation, equipment design is based on minimising pressure drop. At the conclusion of decaffeination the carbon dioxide is unloaded to a holding vessel and both the coffee beans and the adsorbent are unloaded from their respective vessels. The moist decaffeinated beans are then dried to approximately 10% water via a vacuum or hot-air drier. Further processing is the same as for any coffee bean. The supercritical carbon dioxide patent proposes use of moist carbon dioxide between 40°C and 80°C and pressures between 120 and 180 bars. It suggests caffeine removal from the carbon dioxide by lowering the temperature and or pressure to precipitate the caffeine due to lower solubility or to aid adsorption. Product quality is claimed to be improved because of the chemical stability and inertness of carbon dioxide preventing any reactions with the coffee constituents. In addition, supercritical carbon dioxide has a very high selectivity for caffeine, avoiding losses of non-caffeine solids. Product quality is claimed to be comparable to a regular undecaffeinated coffee due to the avoidance of any aroma/flavour precursor loss during the decaffeination process.

5. DECAFFEINATION OF ROASTED COFFEE AND EXTRACT

There are a number of advantages to the decaffeination of roasted coffee extracts.[15] These mainly deal with processing and equipment costs, especially with roasted extract used for the production of instant coffees. Since we are now dealing with the processing of essentially a finished product having good flavour and aroma qualities, most of the process efforts devoted to decaffeination must maintain and preserve these qualities. Usually the first steps include aroma stripping and collection via standard methods. These techniques are well documented in many patents[16] (see also Chapter 5, this volume). A major difficulty in the decaffeination of roasted extracts involves the loss of trace components to the contacting solvent necessary for flavour preservation. Two approaches to solve this problem in recent United States patents involve either the use of an inert, very specific solvent such as liquid or supercritical carbon dioxide for contacting the extract, or the use of a procedure of recovering these trace components by reflux and recycling back to the extract.[17] The latter procedure may also involve maintaining the level of these components in the solvent high enough to prevent extraction from the roasted extract. After decaffeination, processing is similar to the processing of undecaffeinated extract after adding back the stripped aromas.

6. CAFFEINE REFINING

The caffeine removed from coffee is a valuable commodity as a food or drug additive. At the present time the two main sources of caffeine are from coffee as a by-product of decaffeination and from a synthetic reaction between urea and chloroacetic acid. The price of caffeine is governed mainly by the cost of synthesis and demand. To refine coffee-sourced caffeine economically, the purity into a refining system must be in the order of 70% or better. Most of the procedures described above deliver a caffeine of 70 to 85% purity, and a refining system based on this purity was invented by Shuman in 1948. This process combines a number of purifying steps to eliminate various contaminants such as coffee waxes and oils, as well as caffeine-like water-soluble materials that impart a dark colour to a coarsely granular crystal instead of the desired white, needle-shaped crystals of pure caffeine.

The unrefined caffeine derived from the decaffeination process is usually obtained as a water solution or as dried cake. Any residual solvent is steam-distilled and the caffeine sludge that is formed is dissolved with recycled mother liquor from a centrifugation step in the refining process, as shown in Fig. 5. Activated carbon, recycled from a finishing step before final crystallisation, is added to remove most colour-contributing impurities. The slurry is filtered and the spent carbon is discarded or reactivated, or the caffeine may be recovered by use of acetic acid extraction.[18] At this stage the caffeine is 80 to 90% pure. A crystallisation step followed by centrifugation results in two streams. The crystals are redissolved for an additional purification step and the mother liquor is contacted with pure solvent to recover essentially all residual caffeine. Addition of aqueous alkaline solution can substantially aid this separation.[19] The decaffeinated mother liquor is discarded after the trace solvent is recovered. The solvent containing recovered caffeine is sent to an evaporator where it mixes with feed solvent used for decaffeination. Fresh activated carbon is added to the redissolved crystals. After filtration the

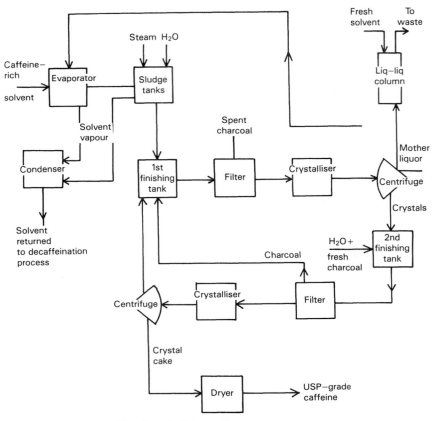

Fig. 5. Caffeine refining process.

carbon is recycled to the first finishing tank. A second crystallisation step followed by centrifugation yields a mother liquor that is recycled to the first finishing tank and a caffeine crystal cake. The cake is air-dried and broken to yield white USP-grade caffeine.

REFERENCES

1. Sivetz, M. and Foote, H. E., *Coffee Processing Technology*, AVI, Westport, Conn., 1963.
2. Meyer, J. F., Roselius, L. and Wimmer, K. H., US Patent No. 897,763, 1 September, 1908.

3. Roselius, W., Vitzthum, O. and Hubert, P., Method for the production of caffeine free coffee extract, US Patent No. 3,843,824, 22 October, 1974.
4. Adler, I. L. and Earle, E. L., Process for preparing a decaffeinated soluble coffee extract, US Patent No. 2,933,395, 19 April, 1960.
5. National Coffee Association, *Newsletter* No. 1862, 4 October, 1982.
6. Patel, J. M., Wolfson, A. B. and Lawrance, B., Decaffeination process, US Patent No. 3,700,464, 24 October, 1972.
7. Mitchell, W. A. and Klose, R., Green coffee decaffeination using edible ester solutions, US Patent No. 3,682,648, 8 August, 1972.
8. Zosel, K., Process for recovering caffeine, US Patent No. 3,806,619, 23 April, 1974.
9. Bichsel, B., *Fd Chem.*, 1979, **4**, 53–62.
10. Horman, I. and Viani, R., *J. Fd Sci.*, 1972, **37**, 925–7.
11. Katz, S. N., Aprotic solvent decaffeination, US Patent No. 4,472,443, 18 September, 1984.
12. Patel, J. M. and Wolfson, A. B., Semi-continuous countercurrent decaffeination process, US Patent No. 3,671,263, 20 June, 1972.
13. Berry, N. E. and Walters, R. H., Process of decaffeinating coffee, US Patent No. 2,309,092, 26 January, 1943.
14. Nutting, L. and Chong, G. S., Continuous process for producing decaffeinated beverage extract, US Patent No. 3,361,571, 2 January, 1968.
15. Prasad, R., Gottesman, M. and Scarella, R. A., Decaffeination of aqueous roasted coffee extract, US Patent No. 4,341,804, 27 July, 1982.
16. Pagliaro, F. A., Franklin, J. G. and Gasser, R. J., Process for the production of a decaffeinated vegetable material and product obtained, Belgium Patent Application No. 835556, 13 May, 1976.
17. Morrison, L. R. Jr, Elder, M. N. and Phillips, J. H., Recovery of noncaffeine solubles in an extract decaffeination process, US Patent No. 4,409,253, 11 October, 1983.
18. Katz, S. N. and Proscia, G. E., Process for recovering caffeine from activated carbon, US Patent No. 4,513,136, 23 April, 1985.
19. Hirsbrunner, P. and Pavillard, B., Purification of caffeine, US Patent No. 4,531,003, 23 July, 1985.

Chapter 4

Roasting and Grinding

R. J. CLARKE*
Formerly of General Foods Ltd, Banbury, Oxon, UK

1. INTRODUCTION

To enable the conversion of the selected green coffee blend into a consumable beverage, three operations are needed: roasting, followed by grinding, and finally brewing. Only by roasting is the characteristic flavour and headspace aroma of coffee developed; and grinding of the roast whole beans is necessary in order that both the soluble solids and flavour volatile substances be sufficiently extracted by infusion or brewing with hot water to provide a beverage of required strength for immediate consumption, or an extract for subsequent drying in instant coffee manufacture.

The roasting and grinding operations are discussed in detail in the following sections with particular reference to the fundamental principles involved.

2. PROCESS FACTORS IN ROASTING

2.1. Mechanisms and Methods

Roasting is a time–temperature-dependent process, whereby chemical changes are induced in the green coffee beans; though marked physical changes in the structure of the coffee are also evident. There is a loss of dry matter, primarily as gaseous carbon dioxide and water (over and above that moisture already present), and other volatile products of the

---

* Present address: Ashby Cottage, Donnington, Chichester, Sussex, UK.

pyrolysis. Roasting is normally carried out under atmospheric conditions with hot combustion gases and excess air as the primary heating agent; though heat may also be provided by contact with hot metal surfaces, solely in more primitive methods, but more generally as a supplement to convection from the hot gases. Some heat may also be supplied by radiation, and in some designs even as the primary source by radiative heating of various kinds.

The roasted whole beans are characterised by the roasting process to which they have been subjected; their so-called degree of roast is reflected in their external colour, the flavour developed, the dry mass loss that has occurred, and the chemical changes in selected components. In simple terms, roast coffee can be described as having a 'light', 'medium' or 'dark' roast; finer gradation terms using these and other parameters are discussed in subsequent sections. Coffees may additionally be described as having been 'fast' roasted (i.e. time of roasting of a few minutes or even less) or 'conventionally' roasted (time in the order of 12–15 min) with many, however, in an intermediate time of 5–8 min.

Degree of roast plays a major part in determining the flavour characteristics of extracts subsequently brewed from roasted coffee, whatever the blend; and speed of roasting is associated with the production of so-called high-yield coffees.

The horizontal rotating drum has been the most usual method of providing contact of hot air in a tumbling coffee bed, but roasters have been designed and are available based on other mechanical principles, as summarised in Fig. 1.

Horizontal rotating drums may provide either batch or continuous roasting whereas the vertical drum and bowl are usually batch. Fluidised-bed roasters generally operate in the batch mode, but the latest designs

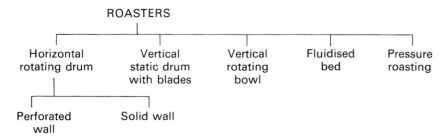

Fig. 1. Mechanical principles in roasting methods.

are often continuous. Further details of actual roasters in these categories are given in section 3.

In addition, all these roasters must provide a cooling facility to bring the roasted coffee immediately after the desired level of roast to ambient temperature, usually by suitable contact with cold air (a process often referred to as 'quenching'). Quenching is a term also applied when water is added by spraying within the roasting section for a similar purpose, but also to provide benefits in grinding (see section 4.1).

A further important feature is the use either of 'once-through' flow of hot gases or of recirculated hot gases, since the latter can have considerable advantages in the rate of heat transfer and overall economy. Both methods are intimately associated also with contemporary needs of reducing the levels of organic volatile substances in the exit gases from the roaster to the stack and finally to atmosphere, which may otherwise be regarded as pollutants subject to legal constraints.

2.2. Chemical Changes

The chemical changes taking place in roasting for each of the individual compounds in green coffee giving rise to the characteristic flavour/aroma of roasted coffee and its brews have been discussed in considerable detail in Volume 1. In processing there are, however, two particular compositional factors of interest.

2.2.1. Dry Matter Roast Loss

The actual dry matter loss observed with different degrees of roast is indicated in Table 1. In practice, accurate measurement of this figure can be difficult as it requires true measurements of the moisture content of both the green and roasted coffee (though arbitrary values can be used). The roasted whole bean may also contain small but significant weights

Table 1
Approximate percentage dry mass loss for different degrees of roast

Degree of roast	Percentage dry mass loss[a]
Light	1–5
Medium	5–8
Dark	8–12
Very dark	>12

[a] On dry basis of green coffee weight.

of entrapped carbon dioxide, which is subsequently released. The total (or as is) mass loss will include the moisture in the original green coffee and is unfortunately the figure very often unhelpfully given in many roasting and related studies. The moisture content is a known variable; generally, though, it is around 12%, but may well fall to about 8% in a dry, warm laboratory.

The actual composition of the gases released from roast and ground coffee, and therefore presumably also from roast whole beans, has been reported as 87% carbon dioxide, 7·3% carbon monoxide and 5·3% nitrogen by volume. Sivetz and Desrosier[1] have presented data for a large continuous roaster, indicating that carbon dioxide constitutes about 8% by volume of the actual gases leaving the roaster, with nitrogen (from combusted and excess air) and oxygen (from the excess air used in combustion) forming the remainder. A number of studies have been carried out on the content (though low) of organic volatile compounds in the roaster gases during roasting. Crouzet and Shin[2] investigated these compounds on condensing and absorbing them from roaster gases for potential reincorporation; and they have also been investigated by others considering their removal as pollutants (see section 2.6). It is not clear what proportion of any particular volatile compound generated is either lost to the roaster gases or retained within the roasted beans, though it is believed that a major proportion of the total volatile substances generated are retained. The roaster gases clearly contain a number of acidic substances in an amount and type not necessarily desired in the roasted coffee.

2.2.2. Carbon Dioxide Evolution and Retention

It is well known that roasted whole beans contain entrapped carbon dioxide, which is retained from that generated (together with evolved water) in the roasting process by pyrolysis. The amount is dependent upon the degree of roast (and also blend) as would be expected, but also upon the method of determination. Immediately after roasting, whole beans contain a quantity of the order of 2–5 ml CO_2 (measured at NTP, 20°C and 760 mmHg pressure), and probably more, per gram of roast coffee. This carbon dioxide is subsequently slowly released, so that the total amount can be assessed by collection methods over substantial periods of time (e.g. some 2500 h), though more rapidly if ground (e.g. some 360 h or less according to the degree of grinding). Barbara[3] has described a more rapid method of collection and reported preliminary figures as high as 10 ml CO_2 per gram of roast coffee. A quantity of 2–

5 ml CO_2 per gram corresponds to only some 0·4–1% of roasted coffee weight, and slightly less of green, which, though significant in many respects, appears to constitute a relatively small proportion of the dry mass lost on roasting. This carbon dioxide must be held under considerable pressure within a roasted bean, which can be calculated to be several atmospheres, e.g.

$$1 \times \frac{\text{Volume of } CO_2}{\text{Void volume of bean}} \text{ per gram; typically } \frac{5 \text{ ml}}{0.78 \text{ ml}} = 6.4 \text{ atm}$$

According to Sivetz and Desrosier,[1] about half of the total carbon dioxide generated is retained in the roast whole bean. For a large-scale continuous roaster, taking 3000 lb (1361 kg) per hour of green coffee and producing 2450 lb (1111 kg) roasted coffee immediately after roasting, these data indicate that the 18·4% total mass loss (on 'as is' green coffee weight basis) at the roaster is made up as follows:

(1) water, 16·7% made up of say 12% initial moisture, with say 4·7% generated moisture from dry matter;
(2) carbon dioxide and other gases, 1·7% loss *during* roasting (50/3000 × 100).

The total dry matter loss on original green coffee weight is therefore (4·7 + 1·7) or 6·4%, corresponding to 7·2% on a dry green coffee weight basis. In addition, a further 1·7% of carbon dioxide is assumed retained in the roasted coffee; so that the total roasting loss when finally assessed on 2400 lb h^{-1} roasted and ground coffee (with virtually total release of carbon dioxide) will be approximately 20% on an as-is green coffee weight. The total possible dry-matter loss on a dry green coffee basis will therefore be as much as 9·1%. The amount of immediately entrapped carbon dioxide on dry roasted coffee weight is $1.7/0.816 = 2.1\%$ w/w, which corresponds to 10 ml CO_2 (at NTP) per gram of coffee, consistent with the data of Barbara.[3] Carbon dioxide release is further discussed in Chapter 7.

A further small contribution to mass or weight loss is the 'chaff' released from either the centrefold of the bean or its surface. Dry processed green robusta beans often contain tenaciously held silverskin, which becomes so-called chaff after roasting. Wet processed arabicas carry much less silverskin and may indeed be polished (see Chapter 2). Few reliable data are available for actual quantities.

The origin of the evolved carbon dioxide (and water) is not clear. Exothermic heat is generated during roasting (see section 2.3); but the reaction involved is not understood, though most certainly involves anhydro formation and other degradative changes in the saccharides present, particularly the sucrose. The latter is of interest in that arabica coffees contain a higher percentage than robusta and this is largely converted on roasting into caramelised substances. However, although the exothermic heat is thought to be lower for robusta, the carbon dioxide content of roasted robusta is in fact believed to be higher than for arabica (at similar roast degree).

2.2.3. Soluble Solids Content

The percentage soluble solids contents of roasted coffees are an often quoted parameter, though a distinction should be drawn between the value arrived at by, say, exhaustive Soxhlet extraction with water at 100°C and that obtained under conditions of some type of household brewing or by an open-pot method, in which, say, 1 litre of boiling water is poured over 53 g ground coffee and stirred for 5 min. Results will differ also according to the degree of grind of the roasted coffee. Reliable data are scarce, though Kroplien[4] provides considerable data using the same extracting method (i.e. 2 g ground coffee simmered with 200 ml water at 100°C for 5 min) for some 36 arabica coffees (calculated average of 31·2%) and seven robusta coffees (calculated average of 30·3%), though of somewhat differing roast colours. In general, it will be true that robusta coffees provide some 3–4% more soluble solids than arabica for the same roast colour; and that the percentage is also greater for darker roasts of the same coffee (though not necessarily when expressed on an original green basis).

Again 'fast' roasted coffees are believed to provide higher percentages of soluble solids than those conventionally roasted; though not necessarily in exhaustive extraction, reflecting a greater availability to extraction of these solids by opening up the bean structure. Maier[5] has provided data for both a robusta and an arabica coffee, showing an increasing yield of water-soluble solids determined by an open-pot method with decreasing roasting time, though there are also differences in percentage roast loss and reflectance colour in the coffees. There is a problem in ensuring that the roasted coffees are exactly comparable in respect to other roasting parameters in such studies.

2.2.4. Other Factors

Excessive temperature in roasting, particularly where the heat transfer is from very hot metal surfaces, is known to cause undesirable chemical changes, especially at the bean surfaces (where the composition is different from that of the body of the bean). 'Charring' and carbonisation in particular affect the bean surfaces, though they may also extend to the whole of the bean. Decreasing air temperatures and decreased proportion of conductive heat is a feature of recirculatory-type roasters (and of others); though the overall quality advantage of hot recirculatory roaster gases has been questioned. Changes occur in these gases outside the intended total conversion of organic volatile materials to carbon dioxide. The kinetics of absorption from these gases by the actual roasting coffee are probably against pick-up of unwanted substances. However, Sivetz and Desrosier[1] discuss the formation of 'tars' depositing on beans (contributing a 'harsh taste') and on the roaster cylinder and other surfaces. Such depositions will occur in 'once-through' roasters, though fluidised-bed roasters (once-through) may well provide cleaner-surfaced roasted beans.

2.3. Heat Factors

After the removal of the original moisture present in the green coffee beans, roasting proper starts at a green bean temperature of about 200°C, after which, through exothermic reactions, escalation of effect readily occurs, requiring considerable control of the process for a given degree of roast. These exothermic reactions can be studied by differential thermal analysis techniques in the laboratory, for example as carried out by Raemy and Lambelet.[6] Figure 2 shows a typical thermal analysis curve for green coffee obtained by the latter authors by plotting heat flow against temperature for a constant heat input. From this diagram, reaction in green arabica coffee appears to start at a temperature as low as 160°C, but not until 210°C (410°F) is reached do these reactions peak, and they fall off at 250°C.

As in other chemical engineering operations, we are interested to know

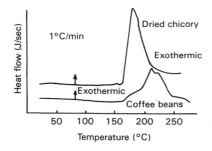

Fig. 2. Calorimetric curves of dried chicory and green coffee beans (both heated in sealed cells). Reproduced from Raemy and Lambelet[6] by permission of the publishers, Blackwell Scientific Publications.

the heat requirements and actual heat balances in roasting, and the various heat transfer factors determining the time required for roasting. These heat factors are considered in the following sections.

2.3.1. Specific Heat of Coffee

Raemy and Lambelet[6] have also reported the specific heats of various coffees from their work in thermal analysis. The specific heat of green coffee is variable depending on its moisture content and also temperature. Assuming a specific heat for water of 1·0 and a law of proportional additivity, it can be assessed from these data that a dry green arabica coffee (Mexico) has a specific heat of 0·395 and a robusta coffee (Togo) a lower figure of 0·32, both at 30°C. At normal moisture contents of around 12%, these figures will therefore be higher (e.g. 0·47 for arabica). Between 20°C and 95°C, the specific heat value will increase by about 30%.

2.3.2. Heat and Power Requirements

A number of calculations have been published for the heat requirements for roasting from actual experimental data taking into account the exothermic heat generated; notably by Vincent et al.[7] and Sivetz and Desrosier.[1] Figures are also available from the manufacturers of roasters. The thermal analysis data of Raemy and Lambelet, already cited, indicated that the measured enthalpy or heat content for a Mexican arabica coffee (7·5% w/w moisture content on green) was between 60 and 100 kcal per kilogram of roasted coffee at roasting (datum temperature not given, presumably ambient). This figure is somewhat lower than would be calculable from the sensible and latent heats needed to bring the green coffee to roasting temperatures. The difference would show the exothermic heat generated. At least this 60–100 kcal figure will be required for actual roasting; in practice, the exiting hot gases have a high enthalpy figure and there are also heat losses from the surfaces of the roasting equipment. Since the required roasting temperature is about 200°C, the exiting hot gases from the roaster cannot be at a lower temperature, and therefore carry heat that is also 'lost' and has to be supplied from the heating source.

Similar enthalpy figures are not given for a robusta coffee and may be different and lower, since although its specific heat is reported as being lower than arabica, the exothermic heat may be less (section 2.2), due to a lower sucrose content. This difference would be consistent with the slower roasting of robusta (but also dry process Brazils).

Vincent et al.[7] determined a so-called energy of roasting from their experimental work on batch-roasting green arabica coffee (Colombian, moisture content not given) in a very small-scale fluidised roaster with 'once-through' flow of hot gases under various conditions. This heat requirement was given by the product of the mass flow of gas (hot air) per unit weight of coffee over the roasting period, the specific heat of the gas, and the drop in gas temperature from its constant inlet to its varying outlet value. The overall effective temperature drop was assessed by integration or graphically. From this the calculated heat requirement figure, which ranged from 119 to 126 kcal per kilogram of *green* coffee for a constant inlet temperature (of 260°C) and constant mass flow rate per unit time (the so-called 'thermal loss' from the roaster), was deducted. This corresponds to the difference between the outlet and inlet temperatures from the roaster (approximately 10°C) after an extended time, up to the actual time of roasting. The rationale for this calculation is not clear; such a figure might equally well, apparently, have been determined from blank control runs without coffee. The 'energy of roasting', or heat required for the actual roasting itself, was then found to be in the range of 87–98 kcal per kilogram of green coffee depending slightly upon the batch weight of coffee; these figures were believed to be close to 'theoretical'. It would be of interest to have the figures based upon roasted coffee weights, corresponding therefore with the enthalpy data of Raemy and Lambelet, and also upon an ambient datum temperature.

Sivetz and Desrosier[1] calculated a figure of 532 Btu per pound of green coffee (about 12% moisture content), or 295 kcal per kilogram, for the actual net heat input requirement from mass and heat balances for roasting in a large continuous drum roaster (3000 lb (1361 kg) per hour) to a total 18·3% weight loss at the roaster. The roasting bean temperature was taken as 196°C and that of the initial hot air 249°C with a datum temperature of 27°C (ambient). The net heat requirement is that after subtraction of the estimated exothermic heat from the gross figure.

A mass and heat balance has to include the following items, preferably on an hourly basis:

(1) weight of dry green coffee entering, together with weight of inherent water;
(2) weight of water generated on roasting;
(3) weight of combustion air (including excess air), and fuel oil used;
(4) weight of gases leaving the roaster stack;
(5) temperatures of the roasting coffee, combustion gases and exit gases;

(6) specific heat of dry coffee, fuel oil, water vapour, combustion air, and of exit gases, for the calculation of sensible heat;
(7) latent heat for the water evaporation, which may be higher for 'bound' water (whether chemically or physically);
(8) exothermic heat input by pyrolysis;
(9) surface heat losses from the roaster, based upon the insulation (i.e. thickness and its thermal conductivity) and internal/external temperatures.

In such a roaster, recirculation of the fuel combustion and roaster gases at a high recycle rate is normally practised; this factor, however, does not essentially affect the heat balance, only the roasting time (see section 2.3.3). Although many of the items above are directly measurable and others deducible for a given degree of roast, the complexity of the process will necessarily lead to only an approximate assessment, even in a continuous process. Application to a batch process causes further calculation difficulties. In the interesting calculation of Sivetz and Desrosier already mentioned, it may be noted that the exothermic heat generated is calculated to provide some 11% of the total gross heat requirement on heating from ambient, or 35·5 kcal per kilogram of green coffee, actual (equivalent to 40·3 and 43 kcal per kilogram on dry green and roasted coffee, respectively). The surface heat losses are about 19% of the net heat requirement, suggesting therefore some 81% heat utilisation for the roasting process itself, a figure very similar to that given by Vincent. The inevitable heat loss from the exit gases (due to their high temperature) is, however, some 25% of the net heat input requirement, a figure not assessed by Vincent. The net heat requirement of 295 kcal per kilogram from ambient temperature is also seen to correspond very closely to that determined from the actual fuel oil usage and its calorific value per kilogram of coffee.

In practice, net heat requirements for roasting are therefore best assessed from fuel-oil consumption data obtainable from manufacturers' literature; thus, for example, 320 kcal per kilogram of green coffee is quoted for a Gothot Rapido-Nova roaster with or without recirculation.

There are, or may be, further heat requirements in a roaster installation. In particular, heat is required to burn or remove organic pollutant substances in the exhaust or exit gases (discussed further in section 2.6). This heat is in the same order as the amount that is required for the roasting process itself, and this information again is most usefully obtained from the manufacturers. For example, Gothot quote a figure of 370 kcal per kilogram of green coffee for combustion in an external afterburner at 450°C, and higher for higher combustion temperatures to give greater removal of pollutants. These figures may, however, be substantially reduced when internal recirculation systems are adopted.

The use of fans, especially those required for recirculation and fluidising purposes, together with any mechanical arrangements for improving hot gas–bean contact, all require electrical energy and need to be taken into account in assessing total energy requirements and therefore running costs. Again figures are obtainable from manufacturers' literature, for example typically 0·0217 kVA per kilogram of green coffee, equating to 18·7 kcal per kilogram, assuming a power factor of 1.

2.3.3. Heat Transfer Rates

Conventional heat transfer factors primarily determine the time actually taken to achieve a given degree of roast (or roast degree in a given time); that is, $Q/t = U \cdot A \cdot \Delta T_m$, where Q is the quantity of heat required in time t (excluding exothermic heat); U is the overall heat transfer coefficient; A is the surface area of the green coffee beans (approximately constant for a given weight); and ΔT_m is the mean temperature differential between the heating air (gases) and the roasting coffee. In practice, the situation is complicated by some variable degree of heat transfer by conduction, and by radiation.

In convective heat transfer, the gas-film heat coefficient component will be important, if not controlling, in the overall coefficient; and therefore the relative movement (velocities) of the beans–hot gases, and similarly that of bean–bean and bean–hot contact surfaces in conduction. An increase in the hot gas to coffee weight ratio, and therefore of effective velocity of gas through the beans, generally produces a marked decrease in roasting time, and/or enables a desirable lowering of inlet gas temperatures, as can be seen in a change from a 'once-through' gas process to a system of recirculation of the hot roaster gases. Increase in gas velocity is also a consequence of the use of fluidised-bed roasting. The gas to coffee weight ratio is therefore a useful parameter for the roasting process, though it strictly needs a qualification of cross-sectional area for gas flow. From the data on a continuous roaster already described, it can be seen that whereas the primary hot air to coffee ratio is only some 0·5:1; the recirculating gas to coffee ratio is of the order of 11 to give a roasting time of some 5–6 min for a gas temperature of about 250°C but actual roasting temperature of 196°C. Sivetz and Desrosier also calculate a value of U (overall heat transfer coefficient), suggesting that this is about 0·5 Btu per hour per square foot of bean surface per degree Fahrenheit air temperature difference (or $1\cdot24\,\text{cal}\,\text{h}^{-1}\,\text{cm}^{-2}\,°\text{C}^{-1}$); they also compare this figure with the thermal conductivity of the coffee beans, assumed to be similar to that of wood, at $1\cdot49\,\text{cal}\,\text{h}^{-1}\,\text{cm}^{-2}$ per ($°\text{C}\,\text{cm}^{-1}$). If the depth of a bean is taken at, say, 0·3 cm, then the contribution to the value of U is indeed high. Although no reliable information appears to be available for the thermal conductivity of coffee beans (at a range of temperatures), the gas-film heat coefficient will be controlling, and efforts to improve this will be worth while.

The data on the fluidised roaster of Vincent *et al.* indicate gas/coffee ratios by calculation in the range 7·3–12·8 to give roasting times of 125–192 s for a light roast, with an inlet temperature of 250°C. In order to cause fluidisation of any particles, certain minimum air velocities must be attained. Both Vincent *et al.* and Sivetz and Desrosier, already cited, have provided calculations for this velocity. During roasting, the roasted beans become lighter in density (to about half their original value), and the air temperature used strongly influences air density, both of which are important in determining the minimum velocity. The former authors find an actual air flow of 45–44 m³ per hour measured at 250°C for green coffee, or 39–37 m³ per hour for roasted coffee, necessary to cause fluidisation over a surface area with a diameter of 100 mm; thus, a fluidising velocity (superficial) of 159–131 cm per second can be calculated for a small bed. The latter authors, on the other hand, calculate a value of 1500 cm per second at 204°C from roasted bean weight and area considerations; which is a true velocity of gas between beans, equivalent to a superficial velocity of about 750 cm per second.

In 'once-through' drum roasters, air to coffee weight ratios are of the order of 3:1 to give roasting times of 10–12 min with relatively high inlet temperatures. Furthermore, the ratio of conductive to convective heat applied has been estimated to be about 1:3 to 2:3 compared with that in a recirculating system at 1:4 to 3:4.

In other designs of roaster, mechanical assistance by means of mixing paddles (Gothot) or a rotating bowl (Probat RZ) to increase heat transfer rates is used, in conjunction with recirculated gases or not. In the rotating bowl roaster, the percentage of conductive heat is stated to be negligible.

2.4. Physical Changes

The most obvious physical change to occur is that of the external colour, ranging from light brown in 'light' roasts to almost black in 'very dark' roasts, together with some progressive exudation of oil to the surface with increasing severity of roast. Uniformity of roast colour at any given degree of roast is generally desired, though this may be frustrated in practice by inherent differences in individual beans, especially of the dry processed type (e.g. immature beans giving rise to 'pales' in the roast—see Chapter 2). The formation of individual charred or carbonised beans at local 'hot spots' is generally to be avoided.

Swelling of beans also progressively occurs, including a 'popping' phase, leading to considerably decreased density, again as a function of the degree of roast, but also of the speed of roasting. Changes in density are reflected in the individual bean densities, or overall bulk densities (free flow or 'packed') of the roasted coffee beans; and especially in the subsequently roasted and ground coffee (of particular significance in its packing, see Chapter 7; and in the filling of commercial percolator columns, see Chapter 5).

Information is available on absolute bean densities of green coffee, which are of the order of $1\cdot25\,\mathrm{g\,ml^{-1}}$ (or $0\cdot80\,\mathrm{ml\,g^{-1}}$), though high-grown arabica are generally higher, $1\cdot30$ (or $0\cdot77\,\mathrm{ml\,g^{-1}}$), which will be close to the values for the dry substance of coffee since the porosity of green beans will be low. Radtke[8] provides some figures for roasted coffees (believed to be arabicas of medium roast); thus, the bean densities are given as $1\cdot428\,\mathrm{ml\,g^{-1}}$ ($0\cdot70\,\mathrm{g\,ml^{-1}}$) for two roast coffees, with somewhat higher specific volumes for three decaffeinated roasted coffees. The 'dry substance' density is given as $0\cdot755\,\mathrm{ml\,g^{-1}}$ ($1\cdot32\,\mathrm{g\,ml^{-1}}$) for the two caffeine-containing roast coffees, corresponding to a porosity (volume of void space/volume of bean) of $0\cdot47$. The degree of swelling on roasting is of the order of 40–60%, i.e. percentage of original density; though robustas swell less than arabicas for the same degree of roast (and are thus said to roast more slowly).

Data are also available on bulk densities; thus, according to Sivetz and Desrosier,[1] a light roast coffee may have a free-flow bulk density of $23\,\mathrm{lb\,ft^{-3}}$ ($369\,\mathrm{kg\,m^{-3}}$) but only $18\,\mathrm{lb\,ft^{-3}}$ ($289\,\mathrm{kg\,m^{-3}}$) when darkly roasted. The fluidised-bed roasted coffees of Vincent et al.[7] had a bulk density of $320-350\,\mathrm{kg\,m^{-3}}$ for a Costa Rica arabica coffee, and $270-330\,\mathrm{kg\,m^{-3}}$ for a Cameroon robusta; in the green form both had a value of 700 (however, the method of measurement is not given).

The internal structure of roasted beans will show differences from that of green coffees; microphotograph comparisons have been published by Radtke.[8] The cell walls are believed to remain intact, but there is a softening of the structure. This is reflected in the variable compressibility of grounds after water extraction, and is of consequence in commercial percolation.

2.5. Measurement of Roast Degree

As already mentioned, degree of roast is qualitatively or visually assessed from external colour. In large-scale commercial practice, a final quantitative assessment is made by colour reflectance readings by a suitable meter, of which there are a number of different types available. It is generally desirable first to take the roasted whole beans, at a suitable time interval after roasting/cooking/quenching, to allow for equilibration, then to grind finely in a clearly specified manner to a particular particle size range, and often to use as the sample for measurement a specified size fraction. Such a sample can also be used for visual comparisons; but it is again advantageous to place the sample in a suitable metal dish and compress (at about 1000 psig or $70 \cdot 3 \, \text{kg cm}^{-2}$) in order to prepare consistently a surface for reliable quantitative measurement of the light reflectance.[1] In this manner, variations of colour within the roasted beans themselves are evened out.

In essence, the various meters available, notably the Photovolt, Gardner, Agtron, EEL and Hunterlab meters (which have been described in references 1, 9 and 10, and elsewhere), rely on the use of photoelectric cells, though there are differences in the nature of the reflected light that is actually measured. For example, the Gardner (Automatic Colour Difference Meter) as described by Smith and White[9] measures the reflected light in terms of three additive parameters, the percentage reflectance, a measure of the deviation from grey in the direction of redness ($+$) or blueness ($-$), by means of tristimulus filters (green, amber and blue) on to three separate photocells which actuate a voltage measuring/indicating device. The initial standardisation is provided by an opaque white tile over the measuring device. The Hunterlab meter (as described in use in US Patent 3,493,389, 1970) also provides results in three parameters; L, a and b. The Photovolt meter also uses a tristimulus filter, but of one colour to a single photocell, providing a single reading which can be referenced to readings from specially prepared coloured tile surfaces with specific and constant reflectance readings.

The reflectance readings of already roast and commercially ground coffee may be determined in a similar manner; that is, by taking a sample which is further ground, and a selected size fraction for actual measurement. It is of interest that the colour reading may not be the same as that of the original roast whole bean coffee, presumably due to changes in storage, which may make the coffee somewhat darker in colour.

While the foregoing methods of determining roast colour are convenient and rapid in process control, other methods are available for assessing roast degree. Accurate in-plant measurement of dry matter mass loss can be used, for example, but involve laboratory checks on moisture contents of green and roasted coffee, even though the methods used may only be giving arbitrary values (see Volume 1, Chapter 2). The measurement of moisture in roasted whole beans will be subject to error, arising from the content of entrapped gases and their contributing weight, as already described. An oven-determined moisture content on roast *whole* beans shortly after roasting will give an elevated value, though this will approximately relate to the roast loss observed at the roaster. After grinding, an oven moisture figure will be closer to the true moisture content, but the actual dry matter loss to be reported on roasting now needs to take into account the weight of released carbon dioxide. Grinding itself may also result in moisture pick-up or loss. As both trigonelline and chlorogenic acids[5] are destroyed according to the degree of roast, laboratory determination of the contents may be used as described by Kwasny[11] and by Pictet and Rehacek.[12] It will generally be necessary to prepare specific correlations for particular roasting methods, plants and blends; and in the case of chlorogenic acid preferably to select and follow a particular chlorogenic acid (e.g. the 4,5-dicaffeoylquinic acid, see Volume 1, Chapter 5). As a guide, the total chlorogenic acids will fall to about 40% of their original value for a medium roast on an equivalent dry green basis.

2.6. Emission Control of Organic Compounds and Chaff

Whereas the odour of gases from a small coffee roaster is generally regarded favourably, such exhaust gases from large roaster installations are regarded as a public nuisance and indeed in some countries as pollutants subject to legal restraint as to the maximum amount of organic material that can be exhausted to the atmosphere. For this reason, the use of so-called afterburners for cleaning has been general for some time. The principle is to recombust these gases at a high temperature in a separate furnace situated in the exhaust stack of the roaster. Such installations consume considerable quantities of heat, much higher than that required by the roaster itself (section 2.3.2). Methods such as wet washing are much less satisfactory.

The emission of organic compounds has been especially studied by members of the Gothot Company, in conjunction with the Gas Heat Institute of Essen in Germany, and their results are reported in reference 13. The measurement of amount was carried out on a total organic basis, specifically as carbon in the units of milligrams of carbon per cubic metre of gas at NTP, or m_n^3. The organic compounds can be trapped on silica gel and estimated or determined by FID (flame ionisation detector) detection on gas samples in a gas chromatograph. By following the content at 1 min intervals during a conventional 6 min batch roast in a Gothot Rapido roaster (without cleaning of exhaust gases except chaff removal), it was found that the total carbon number was quite low at first, but then rose to a maximum of some 1200 mg C m_n^{-3}, as shown in Fig. 3. The average level was 158.

The German Federal Republic law of 1968 stipulated a maximum level of 300 mg m_n^{-3}, though there are expectations that lower limits could be set there and elsewhere. By using a Gothot roaster with gas recirculation (see section 3.2), the effect of different operating temperatures in the furnace was studied. At a temperature of 600°C, the average value was down to 45 (maximum 220), and at 700°C, the average value was further reduced to 15 (maximum 60). With a conventional external afterburner at 450°C, the average value was 54, and it is known that a temperature of 350°C was unsatisfactory. The required temperatures for substantially reducing the carbon level in emissions are therefore not unexpectedly considerably higher than those used in roasting, a problem that this particular Gothot roaster is designed to solve. An alternative solution is the use of catalytic converters which can reduce carbon levels in roaster exhausts at lower temperatures of the order of 350°C. There is little up-to-date published information on this development, except from the brochures of the Barth company.

Similar requirements exist for minimum dust content (specifically chaff

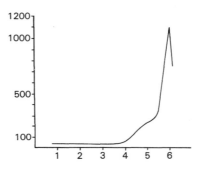

Fig. 3. Characteristic course of emission of volatile compounds (measured as mg C m^{-3} of gases at NTP) with roasting time up to 6 min for a medium roast coffee. Data from reference 13 by courtesy of F. Gothot, GmbH.

from roasting coffee) in exhaust gases, e.g. German Federal law stipulates a limit of 150 mg m_n^{-3}. This problem is simpler to solve, requiring the installation of efficient cyclones in ducting before afterburners or recirculation furnaces.

Emissions and their control from roasters, both batch and continuous, in the US are discussed by Spence.[14]

3. ROASTING EQUIPMENT

The various methods of roasting that have been adopted have been discussed in section 2.1. The commercial plant that is used and available is now briefly described, together with some points of operation. Roasting plant is supplied by various companies specialising in their construction.

3.1. Horizontal Drum Roasters

These roasters are perhaps the oldest and most well established type, particularly those that are batch operated, and are manufactured by such companies as Probat-Werke of Emmerich in Germany, Jabez Burns (Division of Blaw-Knox) of the USA (Neotec in Germany) and G. W. Barth of Ludwigsberg in Germany, in large-scale units with all necessary ancillaries for installation.

There has been a gradual evolution in such roasters from those used in the 19th century, with designs allowing a decreasing proportion of conductive heat from hot metallic surfaces, use of improved furnaces and their siting, and lower hot gas temperatures, through to the use of recirculatory systems.

3.1.1. Batch Drum Roasters

The Probat, Burns 'Thermalo' and Barth 'Tornado' roasters are now characteristic of this type, but differ in that the Probat has a double-walled solid drum so that the hot gases from the furnace (fuel-oil or gas fired) enter through the back end, through the tumbling coffee beans and away to the stack. The four-bag roaster (240 kg batch) is typical of the Probat GO-series with a roasting time of some 10–12 min and turn-round time of 15–18 min. The heating is programmed in two stages, from a controller monitoring the bean temperatures. A high rate of heating is used initially, followed by a lower rate until exothermic reaction occurs, and at a set roasting temperature the furnace is cut off and the roasted beans discharged into a tray with upflow of cold air. The inlet gas

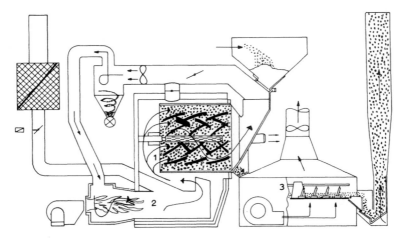

Fig. 4. Probat batch roaster, type R, showing (1) solid-wall roaster drum, (2) furnace, (3) cooling car, and ancillaries. Courtesy of Probat-Werke GmbH.

temperatures will be of the order of 450°C (850°F) or less. This series has been replaced by the R and RR series, which incorporate recirculation of the exiting roaster gases via a cyclone (for the removal of chaff) back to the furnace, but with a return of excess to the roaster stack after cleaning through a catalytic converter or afterburner, as seen in Fig. 4.

The Burns 'Thermalo' roasters differ in that the drum is perforated, allowing the hot gases entering the back end of the drum to pass through the tumbling beans and through the perforations back to the furnace after chaff removal (see Fig. 5).

3.1.2. Continuous Roasters

Burns was the first company to offer a successful continuous roaster (in 1940), again using a perforated drum or cylinder for both the roasting and cooling sections. The cylinder of a 3000 lb h^{-1} roaster rotates at speeds from 2 to 7 rpm, according to the residence time of the beans required, which at the highest speed is some 7 min of which some 5 min will be in the roasting section. The coffee beans are compartmented within the cylinder by means of a spiral, and a central disc separates the heating part from the cooling part of the roaster. This roaster has had a built-in hot gas recirculation system from early on, and with the high gas to coffee ratios the gas temperature need only be of the order of 250°C (500°F). Its construction may be seen in Fig. 6.

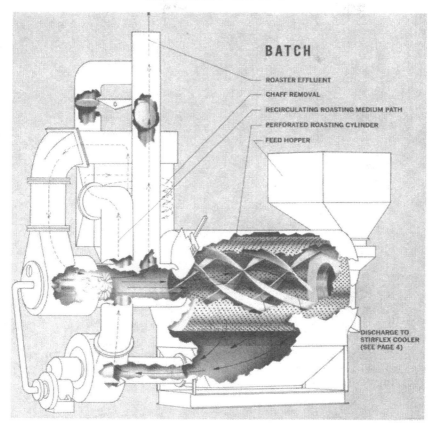

Fig. 5. Burns 'Thermalo' batch roaster, showing perforated roaster cylinder, gas/oil burner with hot gas recirculation and ancillaries, except cooling car. Courtesy of Blaw-Knox Food and Chemical Equipment Company.

Probat introduced a continuous roaster, type RC, in about 1980, the construction of which may be seen in Fig. 7. The roaster again incorporates a recirculation system with chaff removal, together with controlled damper release of a portion of the roaster exit gases to the stack after further cleaning. In this unit, the hot gases are distributed over the tumbling beans by a nozzle distributor running along the horizontal axis of the drum; the cooling section is sited underneath the roasting section. Roasting times are reported to be between 3 and 6 min, again as a consequence of the use of high velocities of the hot gases during roasting.

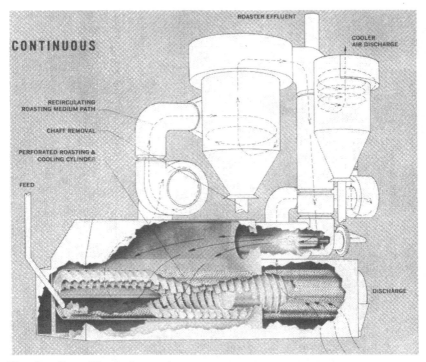

Fig. 6. Burns Thermalo continuous roaster, showing perforated roasting and cooling cylinder, gas/oil burner with hot gas recirculation, and ancillaries. Courtesy of Blaw-Know Food and Chemical Equipment Company.

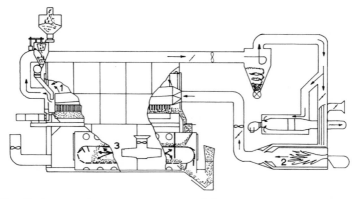

Fig. 7. Probat continuous roaster, type RC, showing (1) roaster cylinder, (2) furnace, (3) cooling section, and ancillaries. By courtesy of Probat-Werke GmbH.

3.2. Vertical Fixed Drum with Paddles

This method of roasting is to be seen in the Gothot Rapido-Nova series of roaster from the Gothot company of Emmerich-Ruhr in Germany. The standard type is illustrated in Fig. 8. A set of paddles to increase the rate of heat transfer between beans and between bean and gas is provided, rotating on a horizontal axis. The exhaust gases are taken through a fan at the top of the vertical cylindrical vessel to cleaning devices. At the end of the roast, discharge is immediate into an integral vessel for cooling. Together with a relatively high gas to coffee ratio, the roasting time is of

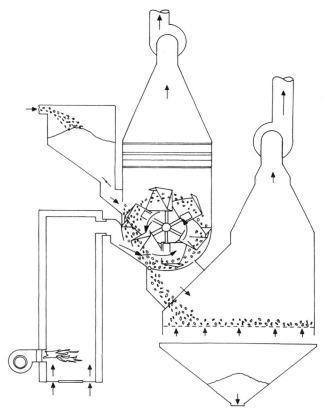

Fig. 8. The Gothot Rapido-Nova standard-type batch roaster, showing fixed roasting drum with paddles, furnace and cooling car. By courtesy of Ferd. Gothot GmbH.

the order of 6 min. This system has been widely patented (e.g. US Patent 3,608,202 in 1971).

This series also now has roasters with in-built recirculation, as illustrated in Fig. 9. The furnace is operated at the fairly high temperature of 600–700°C, so that the exhaust gases (with or without chaff removal) can be almost completely cleaned to a low residual organic level. By controlled introduction of ambient air into the exhaust gases from the furnace, diverted into the roaster chamber, the temperature of roasting can be independently controlled. A further variant allows indirect heat exchange from the exhaust gases to the recirculating air to the furnace. Alternatively,

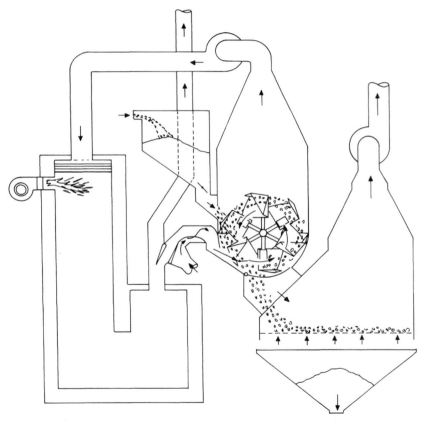

Fig. 9. The Gothot-Nova batch roaster with air recirculation. By courtesy Ferd. Gothot GmbH.

by using a lower combustion temperature, the series includes a roaster with air recirculation but with the use of a catalytic converter in order to clean the exhaust gases.

3.3. Rotating Bowl

A rotating-bowl principle features in the RZ (Radial-Turbo) series of roasters from Probat, which have recently been developed. Coffee beans fed into the centre of a rotating horizontal bowl with a vertical shaft are carried to the periphery of the bowl through centrifugal force assisted by high-temperature air. On reaching a fixed multi-plate ring, they fall back to the centre in spiral-shaped circuits surrounded by hot air. At the end of the roasting, the beans are discharged over the periphery. Figure 10 illustrates the mode of action of this roaster. Short roasting times (about 4–6 min) are employed.

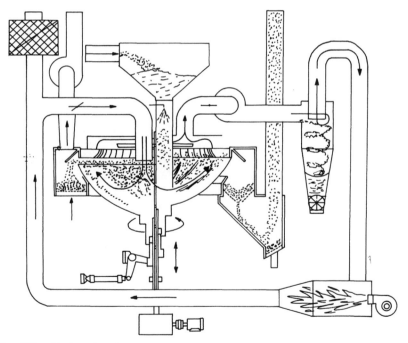

Fig. 10. Probat batch roaster, type RZ, showing the rotating bowl. By courtesy of Probat-Werke GmbH.

3.4. Fluidised Beds

The first commercial fluidised-bed roaster of reasonable size is generally considered to be the Lurgi Aerotherm roaster which was introduced in 1957 and though not now marketed is still in use. The largest batch size is stated to be 110 lb (50 kg), with a roasting time from 2–4 min from 'once-through' hot air indirectly heated.[1,5] The air to coffee weight ratio is calculable at about 10.

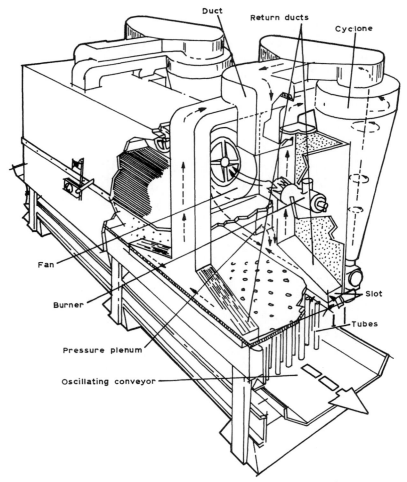

Fig. 11. The Wolverine Jet Zone Roaster, showing the oscillatory conveyor, tubes for hot air and furnace. Reproduced from Ref. 15 by permission of Food Engineering International.

A further fluidised-bed type of roaster is the Wolverine Corporation Jet Zone roaster, which is, however, continuous in operation. Jets of hot air are directed in to the coffee beans fed on to an oscillating conveyor as indicated in Fig. 11. This roaster is described[15] as used for Hills Bros. (California) 'High Yield' coffee.

3.5. Pressure Operation

Sivetz[16] has described the development of these roasters from the Alcalor batch pressure roasters to the Smitherm roaster. He described experimental work with a pilot model of this latter roaster, as patented in US Patent 3,615,688 (by H. L. Smith, 1971). This model is continuously operated using nitrogen gas pressure, in a chamber designed for 150 psig pressure and 440°F (227°C) temperature and fitted with a recirculating gas system. The roasting time is given as 5·33 min, and the ratio of hot gases to coffee is reported to be 3500 cfm (at 80 psig) to 6 lb coffee per minute, suggesting that the weight ratio is 70:1. The reported data by comparison with conventional atmospheric roasting (sample roaster) suggested, as expected, that these pressure roastings for the same reflectance roast colour gave roasted coffee which on brewing had a higher acidity, somewhat higher yield of soluble solids and somewhat higher bulk density.

3.6. Roaster Ancillaries

A complete roaster installation consists of more than the roaster(s) themselves. For batch roasters, batching must be carried out with green coffees drawn from the silos, with either weighing or volumetric input from a control panel. The cooled roasted coffee has (assuming in-built cooling/quenching arrangements) to be conveyed to storage vessels or silos awaiting the next operation of grinding (as required), usually via a destoner. It is useful at this stage also to be able to weigh the roasted coffee. The need for afterburners (if not in-built) and chaff removal devices has already been mentioned.

4. PROCESS FACTORS IN GRINDING

4.1. Mechanism of Grinding

Roasted whole beans require a cutting rather than a tearing or crushing action to provide a ground coffee with particles of suitable size and shape for subsequent brewing, and this is best produced by means of cutting rolls of various designs. Nevertheless, various impact-type grinders have been used for producing finer grind coffee, and recently so-called flaking rolls have been used to produce particles of a flaked shape, which, it is claimed, gives benefits on brewing.

Clearly the finer the average particle size, the more rapid will be the rate of extraction by water of soluble solids and volatile components of the coffee. In practice, the method of brewing and equipment used will dictate the degree of grinding which is desirable, as discussed in this chapter and in Chapter 8. This is equally true in the process of grinding coffee subsequently to be extracted in large-scale plant for instant coffee manufacture. There can be a problem in home percolators with 'fines'. This is negligible with filters, but the coffee bed should not consist of such fine particles as to impede percolation.

Coffee grinds are qualitatively defined as being 'coarse', 'medium' or 'fine', or especially as in US usage according to the brewing device for which they are intended, e.g. 'percolator', 'drip-pot', 'filter', 'vacuum makers' or 'espresso'. There is no consensus of agreement, whether national or international, as to the average particle size (and range) that constitutes required sizes in this nomenclature; but various recommendations are available for the various grinds in terms of screen analyses.[17] 'Coarse' or 'medium' grinds are used for household percolators (and generally somewhat coarser for large-scale percolators); whereas 'fine' grinds are used in filter devices, and often somewhat finer in espresso machines. An indication of the average size of particles in grinds typically used is given in Table 2, where it is evident that US grinds tend to be coarser for each named grind than in Europe.

Achieving as uniform a grind as possible is important, as discussed and defined in the next section. Uniformity is generally best obtained by multistage grinding devices (i.e. with a series of grinding rolls, progressively reducing particle size). Uniformity of grinding is also dependent upon the condition of brittleness of the roasted coffee, its moisture content and degree of roast. Grindability is improved by the prior addition of water at an optimum quantity of around 7%, as especially practised in ground coffee for subsequent large-scale percolation.

Table 2
Average particle size (μm) in coffee grinds[18]

Brewing device	Europe	USA
Percolator	850	1130
Filter	430	800

Really fine grinding (say to an average particle size of about 50 µm or less) is never really required for home brewing or large-scale percolation, though the requirements of Turkish-type coffee, and of specific processes and products in instant coffee manufacture, e.g. British Patent 2,006,603B (General Foods, 1982) need to be met. Such grinding is best carried out under cryogenic conditions (i.e. at sub-zero temperatures by use of solid carbon dioxide or liquid nitrogen), especially using impact-type mills as described in a number of patents.

Cryogenic grinding has also been recommended for maximising the retention of flavour volatiles within the roast and ground coffee, which are otherwise released in conventional grinding, especially when accompanied by the generation of heat. However, so-called grinder gas is mentioned in a number of patents (see Chapter 5). This may be collected by means of condensing devices attached at a suitable position in the exit passages of the grinder. Such condensates will contain useful quantities of aroma/flavour volatiles, which may be incorporated later in the process of manufacture and packing. The major part of these gases is, however, the residual carbon dioxide in the roast whole beans (itself a variable amount, see this chapter, section 2.2.2) and water vapour. Immediately on grinding, depending upon the degree of grind, about half of this residual carbon dioxide is released, causing some potential hazard if not properly vented. The remainder of the carbon dioxide still present in the ground coffee is subsequently slowly released. This release has been studied by a number of workers, as discussed in Chapter 7.

The grinding of coffee will also release a further quantity of chaff, either that which is not released from the surface on roasting or that remaining in the centrefold of the whole beans. This material can be disposed of, but is often incorporated with the ground coffee in a so-called 'normaliser', by mixing under the action of mixing blades. Particularly with dark roast coffees, this process is advantageous in absorbing any exuding oil on the surface and in keeping the ground coffee reasonably free-flowing in subsequent handling. 'Normalising' has a further function in that it is a means of controlling the bulk density (see section 4.3), though only within certain limits, by variation of blade speed and disposition.

4.2. Size Analysis

The traditional method of assessing degree of grind is from the results of sieving analyses using a number (or nest) of different screen or sieve mesh sizes (of woven wire mesh with square apertures of defined dimensions), in a specified procedure. Some four or five screen mesh sizes may be used at a time, though only two are really needed to define adequately a given degree of grind.

4.2.1. Screen Sizes

Considerable rationalisation in screen mesh sizes has taken place over the years through the activities of the International Standards Organization and the adoption of its recommendations by national standards bodies. For woven wire mesh screens, as would be used for the screen analysis of roast and ground coffee (also instant coffee), the relevant ISO standard relating to aperture size is ISO 565 (1983). The square apertures of the recommended screens in a set follow sequences of dimensions based upon so-called 'preferred' numbers as set out in ISO 3 and also ISO 497. Three series are defined in ISO 565: first, a principal series and then two supplementary series. It is necessary to consult all these standards to have a full understanding of the scheme. There are three main sets of preferred numbers of interest in screening: the full R.40 series, such that successive screen apertures are in the ratio $\sqrt[40]{10} = 1\cdot059$; the full R.20 series, in the ratio of $\sqrt[20]{10} = 1\cdot1220$, and the R.10 series, $\sqrt[10]{10} = 1\cdot258$. In practice, there is some rounding-off of actual aperture dimensions. Two subdivisions are also made, R.20/3 and R.40/3, which make a narrower selection, e.g. R.20/3 every fourth size in sequence from the full R.40 series. The aperture sizes of the R.10 and R.20 series will also be found within the full R.40 series. The selection of screen apertures that has been made for ISO 565 is for the principal sizes, the R.20/3 series (approximate ratio, 1·42); and for the two supplementary sizes, the remaining aperture sizes from the R.20 series and separately those from the R.40/3 series.

The various national standards, such as those of the British Standards Institution and of various bodies in the USA (e.g. those designating Tyler and ASTM screens), in the past designated screen sizes by use of numbers, though with clearly defined aperture dimensions (in millimetres, microns or inches). Now, however, in the UK (1976), France (1970), Germany (1977), USSR (1973), Italy (1973) and Japan (1982), the individual institutions have standardised on aperture size designations (expressed in millimetres or microns), with sizes again both in a principal and supplementary series. However, the selections made differ in the different countries and differ from that of ISO 565. Thus, for example, in the UK (BS 410, 1976) the principal and recommended series is based upon R.20/3 (i.e. as IS 565), but with one supplementary set, from R.40/3; so that the screen apertures in the total list have an approximate sequence ratio of 1·2 (approximately the fourth root of two); similarly now in the USA (ASTM, E-11-81, 1981) with one listing. Germany and France, however, make use of R.10 as the principal series and R.20 as a supplementary series. Some confusion may inevitably result, though the selections made can be clearly seen and compared as set out in a French National Standard (AFNOR X11–508, 1983). The USA have, in addition to the ASTM series, the Tyler series, much used for roast and ground coffee, which is numbered according to the number of nominal meshes per linear inch.

The important specification of other characteristic dimensions of woven wire mesh screens (e.g. wire diameter) is given in ISO 3310/1, and nationally usually within the same standards as cited above. There are other ISO standards relating to vocabulary (ISO 2395) and procedures (ISO 2591). Round sieves are generally standardised at 8 in diameter, or 200 mm.

4.2.2. Screen Analysis Results

It is not possible to grind coffee to a single uniform particle size, and a distribution around an average is to be expected. The results of a size analysis, based upon use of the screen sizes described above, are most conveniently plotted graphically with the screen mesh size (i.e. aperture in microns) on a linear scale on the y-axis and the weight cumulative oversize on the x-axis. Such a plot is shown in Fig. 12, taken from tabulated data on three typical size analyses from sieving given in BS 3999 Part 8, 1982 (Performance of Electric Household Coffee Makers). The average particle size may be approximately assessed as that of the aperture at which there is a 50% cumulative oversize. Such plots are generally found to be quite linear, except at the extremes. The slope of the line indicates the degree of uniformity of particle size, a parameter

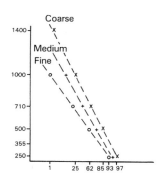

Fig. 12. Screen analyses for different degrees of grind of roast and ground coffee, plotting percentage cumulative amount by weight held at each screen size (aperture in μm) according to the R.20/3 series of screens (IS 565 or BS 410). x---x, Coarse; +---+, medium; ○---○, fine. Data from tabulation in BS 3999 Part 8 (1982).

quantitatively assessable as a coefficient of variation (CV). The CV value is obtained by determining the difference between the apertures for oversize amounts of 16% and 50% and dividing by the aperture for 50% retention, and expressing this ratio as a percentage figure. Single-stage grinding (i.e. once through a single grinding device) will show a relatively high CV figure; but multi-stage grinding will be found to give lower CV values, typically 30% for roasted coffee grinds. The use of the R.20/3 series of screens gives a satisfactory separation between aperture dimensions (i.e. sequential ratio of about 1.41) for a reasonable spread of retained weights on the screens. Tolerances of aperture sizes in manufacture are of importance; BS 410 shows an allowable tolerance on the average size of apertures of about $\pm 3 \cdot 2\%$ on each screen. Graphs such as Fig. 12 can be used to assess the effect of using screens at the extremes of the tolerances allowed, which is relatively small except at the larger sieve sizes. For reliable results in test sieving, there is a need to use test screens which are certifiable to the specifications set.

In 1958 Lockhart,[19] at the former Coffee Brewing Institute in New York City, studied coffee grinds by sieve analyses of various commercial blends available at the time. However, it would have been more satisfactory if he had plotted aperture size on a linear scale as used in Fig. 12 rather than on a logarithmic scale.

Grind standards are often given in terms of the screen analyses themselves (see reference 17). Table 3 shows recommended grinds in the USA (by the US Department of Commerce), which, though established

Table 3
Screen analyses for recommended coffee grinds

Grind description	Combined amount retained on:		Amount passing through No. 28 Tyler screen (%)	Approximate average particle size[e] (μm)
	No. 10[a] + 14[b] Tyler screens (%)	No. 20[c] + 28[d] Tyler screens (%)		
Regular	33	55	12 ± 3	1130
Drip	7	73	20 ± 4	800
Fine	0	70	30 + 10−5	680

[a] Equivalent to 12 US mesh (1680 μm, now 1·70 mm).
[b] Equivalent to 16 US mesh (1190 μm, now 1·18 mm).
[c] Equivalent to 20 US mesh (840 μm, now 850 μm).
[d] Equivalent to 30 US mesh (590 μm, now 600 μm).
[e] Computed by graphical plotting for screen size at 50% cumulative weight.

as long ago as 1948, still appear generally acceptable. Some screen apertures used are slightly different in dimensions from corresponding US (1981) and international standards.

The screen analyses of Table 3 are to be conducted with a Rotap machine, which imparts a circular and tapping motion (fixed in character) to the set of sieves, originally manufactured by the W. S. Tyler company. The machine incorporates a timer to cut off the motion, at 5 min for the above analyses. A 100 g sample of the roast and ground coffee is taken for the test. Other types of sieve 'shaker' are available; BS 5688 Part 26 (1981) (although not specifically for ground coffee) recommends a particular shaking device, defining the required circular and tapping motion.

It will be evident that screen analyses will give slightly different results depending upon the mode and degree of vibration applied and other procedural factors in the screen analysis, e.g. those causing abrasion and attrition, and agglomeration of particles; and also the 'blinding' of screens by oily, non-free-flowing coffee particles. The latter should be overcome by the vibration applied (i.e. vertical shaking or rotary motion or both). Repeatability and reproducibility of results are important, though by strict adherence to specified procedures there will be little problem with the coarser grinds. The advent of rather finer grinds than those in Table 3 has prompted the use of screen analysis techniques with more efficient 'deblinding' devices. The air-jet sieve for this purpose has been described by Van Veen and Vreeswijk,[20] and consists of a fixed screen in a closed sieve house in which there is a slight applied vacuum; there is also a rotating air slit just under the screen. The air is blown through the screen and then away through it carrying the fine particles to a collector, leaving particles on the screen corresponding to that screen size which are removed and weighed. Such an arrangement will of course only give one screen analysis figure at a time; and further values for different-sized screens are obtained from a given sample representatively subdivided into portions for test. Reference 20 describes the various parameters in the procedure for optimal results and recommends the required loading (15 g coffee), the sieving time (4 min) and the pressure difference (400 mmHg). Results are plotted on a linear-probability graph as previously discussed, and are compared with those of conventional screen analysis procedures (vibration). In roast and ground coffees of about 400 μm particle size average, it is evident that though both methods give similar sloped plots, the air-jet method gives results some 50 μm smaller for the same cumulative weight percentage, and is to be recommended.

Deterioration of grinder performance resulting from wear, bad settings or lack of maintenance can be readily detected by screen analyses, e.g. changes of slope in linear plots and excess formation of fines.

4.2.3. Alternative Particle Size Measurement

The average particle size from screen analysis can be computed by more sophisticated statistical interpretation. Actual particle sizes can now be rapidly determined, together with amounts at a given particle size by use of sizing laser beams. This is a feature of a number of commercial instruments now available, though little has been published apart from the brochures of the manufacturers (e.g. Malvern Instruments).

4.3. Bulk Density of Ground Coffee

The bulk density of ground coffee can be assessed in one of two ways, free fall or packed measurement. In the former the weight of coffee required to fill a container in free fall, under specified conditions and after levelling the surface, is determined; and in the latter the weight of coffee to just fill a similar container after applying vibration or 'jogging' in a clearly defined manner is determined. Such methods are similar to those used for instant coffee, as discussed in Chapter 6.

Different values will be found according to blend, degree of roast (decreasing with increasing severity), grinding degree (increasing with increasing fineness), moisture content and treatment in a normaliser; values will of course increase from free fall to packed measurements (about 15–20% increase). These values will be related to the originating whole roast beans, and to the weight to volume ratios experienced in actual packing, which need to be assessed directly in practice. A fine grind coffee may have a bulk density of 0·39–0·47 g per ml in free fall.

5. GRINDING EQUIPMENT

For small-scale grinding, variants of the disc mill are commonly used. The discs, one of which rotates, carry serrated surfaces and are separated by an adjustable gap to provide the degree of grind required. Feeding is at the centre, and discharge is from the periphery of the discs. For large-scale grinding (i.e. of the order of 1000 kg roast and ground coffee per hour), the multi-roll mill is almost entirely used, the history of which, together with many details of construction and operation, is described by Sivetz and Desrosier.[1] The type of serration on the rolls is important; the

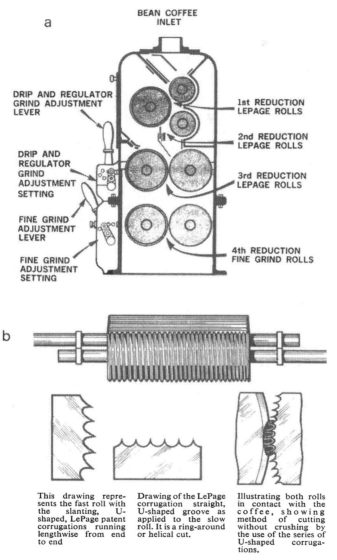

Fig. 13. (a) Gump Granulizer roller assembly and (b) Lepage cutting rolls. Courtesy of Blaw-Knox Food and Chemical Equipment Company.

original patented invention of Lepage for roast coffee provides for slanting U-shaped corrugations running lengthwise in the fast roll and a straight U-shaped corrugated ring around the slow roll. A pair of rolls run at differential speeds with adjustable gaps between them. In the Gump grinder using these rolls in a number of models at different capacities, the housing carries three pairs of rolls, enabling reduction down to the 'drip' grind of Table 3 (Fig. 13); an additional pair of rolls is required for fine grinding. The Probat-Werke company also manufactures a multi-roll grinder with a number of special features.

'Normalisers' can be fitted to the multi-roll grinders so that the chaff released can be incorporated with the roast and ground coffee by means of a trough with mixing blades. The combination of 'grinding' and 'normalising' is commonly known to take place in a 'Granulizer'. The normalising action has an advantage in that since it is also adjustable it can be used to control bulk density of the roast and ground coffee, i.e. usually towards higher figures which may be necessary in subsequent packing.

For intermediate capacity grinding for the finer grinds, single-passage grinding is used (though classification and recycle are possible). The Fitzmill mill is available, which has vertically rotating blades of various designs (cutting and hammer) within a housing with a perforated screen at the base (of different available hole sizes) through which the ground coffee passes to a collecting container. The Entoleter of Henry Simon is another type of mill that is used (also for killing insects in wheat) and is based on a high-velocity impact principle.

REFERENCES

1. Sivetz, M. and Desrosier, N. W., *Coffee Technology*, AVI, Westport, Conn., 1979.
2. Crouzet, J. and Shin, E. K., *Café Cacao Thé*, 1981, **25**, 127–36.
3. Barbara, C. E., *Proc. 3rd Coll. ASIC*, 1967, 436–42.
4. Kroplien, U., *Green and Roasted Coffee Tests*, Gordian-Max Rieck, Hamburg, 1963.
5. Maier, H. G., *Proc. 11th Coll. ASIC*, 1984, 291–6.
6. Raemy, A. and Lambelet, P., *J. Fd Technol. (UK)*, 1982, **17**, 451–60.
7. Vincent, J. C., Aronja, J-L., Rios, G., Gilbert, H. and Roche, G., *Proc. 8th Coll. ASIC*, 1977, 217–26.
8. Radtke, R., *Proc. 7th Coll. ASIC*, 1975, 323–33.
9. Smith, R. F. and White, G. W., *Proc. 2nd Coll. ASIC*, 1965, 207–11.
10. Paardekooper, E. J. C., Driessen, J. and Cornelissen, J., *Proc. 4th Coll. ASIC*, 1969, 131–9.

11. Kwasny, H., *Lebensm. chem. gerichtl. Chem.*, 1978, **32**, 36–8.
12. Pictet, G. and Rehacek, J., *Proc. 10th Coll. ASIC*, 1980, 219–34.
13. Anon, *Coffee Int.*, 1975, **1**, 36–7.
14. Spence, B., *World Coffee and Tea*, 1972, June, 56–60.
15. Anon, *Fd Engng Int.*, 1979, January, 24–5.
16. Sivetz, M., *Proc. 6th Coll. ASIC*, 1973, 199–221.
17. Clarke, R. J., *Food Proc. Marketing*, 1965, **26**, 9–14.
18. Clarke, R. J., *Proc. 11th Coll. ASIC*, 1985, 281–90.
19. Lockhart, E. E., *Fd Res.*, 1958, **24**(1), 91–6.
20. Van Veen, Th. and Vreeswijk, J. H., *Proc. 4th Coll. ASIC*, 1969, 192–200.

Chapter 5

Extraction

R. J. CLARKE*

Formerly of General Foods Ltd, Banbury, Oxon, UK

1. INTRODUCTION

After roasting and grinding, extraction is the key operation in the large-scale manufacture of instant coffee, in which both soluble solids and volatile aroma/flavour compounds are extracted. The direct brewing or infusion by hot water of roast coffee in the home to provide an immediate beverage for drinking is separately discussed in Chapter 8. Large-scale extraction has a number of distinctive features, consequent upon the problems of scale, productivity and economic factors. It is necessary furthermore to provide extracts that can subsequently be dried satisfactorily (as described in Chapter 6) to a convenient instant product, which is also stable and of good flavour on making up to a drinking concentration of solubles with hot water.

The extraction of roast and ground coffee is, in fact, a highly complex operation, the scientific fundamentals of which are very difficult to unravel. This situation is reflected in the complex but still unsatisfactory mathematical equations and procedures needed to model coffee extraction systems fully. The 'mini-engineering' of home brewing presents a similar picture when studied in depth; but large-scale extraction has many further elements not least the fact that the operation is not isothermal. Extraction is often referred to as 'leaching' in texts of chemical engineering (see reference 1), and although there is a reasonable corpus of knowledge, it is only recently that the subject has been more fully studied, in particular

* Present address: Ashby Cottage, Donnington, Chichester, Sussex, UK.

by Schwartzberg[2] and Spanninks,[3] who both include in their frame of reference a wide range of different contacting systems for the solvent and solid. Liquid water is the only solvent used in practice for extracting coffee soluble solids; however, it also extracts the volatile components present in small quantities. Organic solvents are not used in practice for extracting volatile compounds or soluble solids; though there are many patented processes, for example the use of inert gases (whether supercritical or not), for the volatile components.

It is not surprising that the extraction of roasted coffee in practice has many elements of craft, skill and experience, though various mathematical correlations between the more obvious variables have proved to be useful guides. In optimising the operation for extract quality, yield from coffee, concentration of extract and productivity factors, numerous variants of procedures have been described largely in the patent literature; and these include, particularly in the last decade, specialised procedures for separate handling and incorporation of the important volatile components in the finished product.

2. MECHANISMS AND METHODS

2.1. Methods

The necessary contact between any solid extractable foodstuff and its solvent, resulting in separation of the solute in solution, may be arranged by a number of different methods as described in various texts.[1-4] In the extraction of roasted coffee with water, there are three main methods, though one of these is the most commonly used. Unlike single-stage home brewing, providing very dilute coffee solutions (c. 1% w/w), it is necessary to arrange multi-stage or continuous countercurrent operation in order to obtain efficient extraction with an adequately high concentration of soluble solids (15–25% w/w) in the final extract taken. Such a concentration is necessary to reduce the amount of associated water to be subsequently handled and removed to provide a dry product, primarily for economic reasons. Types of extraction plant for coffee may be subdivided as in Fig. 1,

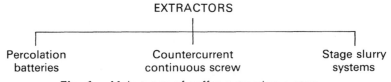

Fig. 1. Main types of coffee extraction system.

Percolation batteries, which are the most commonly used system, derive from the 'Shanks' or diffusion batteries, as also used for sugar extraction from sugar beets or cane. The roast and ground coffee is held as a static bed in vessels (or vertical 'columns') with internal separation of liquor from one stage or column to the next. The flow of hot water to the roasted coffee is countercurrent, i.e. the feed is continuous to the column of the most exhausted coffee; though the draw-off of extract from the so-called 'fresh' column is intermittent. As each column is exhausted, it is isolated from the battery and the spent grounds are discharged, so it is replaced in the battery by a 'fresh' column, i.e. the discharged column is refilled with fresh coffee and replaced 'on stream'. The sequence of events may be followed in Fig. 2. The number of columns in a battery may range from five to eight. Such a system can be used with a feed water temperature of 100°C, though good insulation or other means would be needed to maintain the same temperature or similar throughout the battery. In industrial practice, higher feed water temperatures up to 180°C are used, when, by natural heat loss from the column, the temperature of the liquor and grounds in each column will progressively fall until the liquor contacting the freshest coffee will be at around 100°C. A temperature profile is therefore established over the battery, which also may be artificially arranged by use of heaters and/or coolers placed in stream between particular columns as required. The use of higher feed water temperatures enables higher yields to be taken (see section 2.2), and also higher concentrations of extract. A consequence of the use of higher temperatures is that the system must be kept under pressure to maintain hydraulic conditions (e.g. water at 175°C requires 120 psig, or 828 kPa, overpressure), and that the individual columns and their associated pipework must be designed for operating pressures in considerable excess of the hydraulic minimum. There is, therefore, also a pressure profile over the battery; it can be artificially arranged by throttling the flow at the valve before the fresh column to have atmospheric pressure or thereabouts in this column, while the maximum pressure will be shown at the outlet from the feed pump. Two key factors in operation are the cycle time (assessed by the time difference between placing the fresh column on stream and the finish of drawing off extract from that column) since this time determines other factors, notably productivity, see section 2.5.3; and the weight of extract drawn off per cycle and its soluble solids concentration, determining the yield obtained (see section 2.5.2). For any given percolation battery, the grind size (Chapter 4) has to be decided, since although finer grinds will increase the rate of extraction as expected and thus decrease the cycle time, their use may be self-defeating in that the

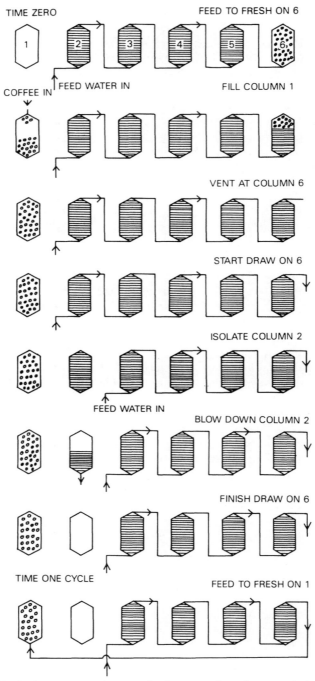

Fig. 2. Typical sequence of events in the operation of a percolation battery for instant coffee. Shape, size and number of columns are diagrammatic.

higher pressure drop across the battery cannot be sustained. It is generally true that larger installations (capacities) require coarser grinds, not least from surface/volume considerations of the column in physically supporting the bed of coffee against compression and collapse.

Such a battery will inevitably have many possible variants of operation, and a selection of patented processes, some now very old, is given in Table 1.

Table 1
Some patented modes of operation of coffee percolation batteries

Mode of operation	Patent No./date	Assignee
1. Temperature profiling, two columns as the 'cold' and four as the 'hot' section	USP 2,915,399 (1959)	General Foods Corporation
2. Temperature profiling—use of inter-column heating	USP 3,655,398 (1972)	General Foods Corporation
3. Optimising conditions	USP 3,700,463 (1972)	Procter & Gamble
4. Use of inter-column evaporation and make-up of feed to fresh column	USP 4,129,665; BP 1,537,205 (1978)	Nestlé
5. Use of two fresh stage columns and split draw-off	BP 1,547,242 (1979)	Nestlé

The 'split draw-off' concept is frequently mentioned, in which the second part of a draw is evaporated to a higher concentration and then added to the first part of the draw before drying. A truly countercurrent system can be established within a single inclined or horizontal cylindrically shaped vessel, by movement of the coffee by a rotating screw conveyor (single or twin) against the flowing hot water. Such a system is again available and now widely used for sugar beet extraction. There have been a number of designs and commercial models for roast coffee extraction, where again it is necessary to arrange a temperature profile and operation under pressure. The production of extract is now continuous. Variants of operation are again possible, including the concept of split draw as described in section 3.2.

So-called 'slurry' extraction is obtained by the use of tanks or vessels (fitted with agitators) for contacting water with coffee, and separating devices such as centrifuges between the contacting stages, of which there would need to be at least three to obtain an adequate concentration of soluble solids in extract in countercurrent flow. Slurry extraction does enable the use of relatively fine grinds for the roasted coffee (unlike the two foregoing methods), but large-scale plant for a similar capacity will be rather more expensive in capital and running costs. There are a number of patents describing such processes, e.g. US Patent 3,529,968 (Procter & Gamble, 1970).

2.2. Mechanism of Soluble Solids Extraction

Roasted coffee differs in a number of respects from other foodstuff solids which are extracted. It is a cellular substance with a matrix of carbohydrate and oil, and can become compressible, especially when higher yields are taken. Historically also, in extracting at lower yields such as those obtainable at 100°C, extracts were obtained which were difficult to dry subsequently to free-flowing and 'non-tacky' powders, so that corn syrup solids or other similar carbohydrate materials were often added (up to about 50% of solids weight). It was then realised that roast coffee itself could provide such beneficial materials in sufficient amount, by use of extracting water at higher temperatures up to 175°C, as described by Morgenthaler of Nestlé in 1943 (US Patent 2,324,526). The first large-scale use of such elevated temperatures was developed about 1950. For this a percolation battery operable under pressure was particularly suitable, and made possible the sale by Maxwell House first in the USA of 100% pure instant coffee.

It is probable that the extraction of roasted coffee is not a completely physical operation, even at 100°C, with some chemical transformations taking place, i.e. particularly some cleavage of large molecular mass polysaccharides resulting in their solubilisation, as described in section 2.2.2. Release of some components held to cell walls, even by covalent bonds, may take place, such as some of the arabinogalactan (see Volume 1, Chapter 3).

Nevertheless, extraction can be treated as a unit operation of chemical engineering,[5,6] in which we have to consider the mass transfer of diffusible components from inside a solid phase to a bulk liquid phase, as described in section 2.5.2. Both sugar cane/beet and oil seeds (such as soya) are also cellular substances, extracted by hot water (though not elevated pressure) and by organic solvents, respectively. It is of interest that both these materials have to be prepared for extraction; sugar beets are cut into 'cosettes' to minimise diffusion distance without excessive opening of cell walls which perform as a selective membrane preventing release of high molecular weight compounds (regarded here as an impurity in the aqueous sucrose solution). Soya beans need to be flaked to facilitate extraction. With roast coffee, grinding is an important factor in successful extraction, though 'flaking' has been proposed for coffee to be brewed, e.g. US Patent 3,652,292 (Procter & Gamble, 1972).

2.3. Mechanism of Volatile Compound Extraction

It is equally if not more important in roast coffee extraction to ensure the extraction and avoid the loss of volatile compounds, of which many, but not all, will provide the desired flavour/aroma characteristics of coffee, as discussed in Volume 1, Chapter 8. It is still not clear as to which compounds and in what amounts relative to each other are significant for coffee flavour. However, coffee flavour is not a single entity, as variants of blend and of roast colour will indicate.

The volatile compounds (of which over 700 different compounds have now been found to be present) comprise a whole range of different compounds of very different physical properties. They are present in only small amount (excluding some semi-volatile acids), i.e. $c.$ 0·1% total, with some important compounds present at levels of only parts per billion. They are likely to be distributed across the oil and the carbohydrate matrix of the roasted coffee, with many compounds self-evidently present in the oil (i.e. the aroma of mechanically expressed coffee oil).

The extraction systems discussed for extracting the soluble solids will also extract the volatile compounds. Although the amount of individual compounds present will be well below their maximum solubility in the amount of water used, some non-polar compounds, preferring an oil phase, will be more difficult to extract by water than others, as discussed in section 2.5.2.

Often in commercial practice it is decided to pre-extract the volatile compounds by means of steam stripping before conventional water extraction; and in other cases to strip out the volatile substances from the extract before bulk evaporation to a higher soluble solids concentration, reincorporating either or both in the concentrated extract before drying. The mechanism of stripping is again well recognised in chemical engineering and discussed in section 2.6. As a further variant, coffee oil expressed from roast coffee may be used as a source of headspace aroma volatiles, or, from spent grounds coffee, as a vehicle for added aromas/flavours; both oils are then plated on to the final instant coffee product.

Table 2
Approximate typical composition of a medium roast coffee (arabica)

Compounds	% Dry basis	Comments
1. *Alkaloids* (caffeine)	1·4	Small amount of loss on roasting (*c.* 4% rel.). Overall figure slightly increased from green
2. *Trigonelline*	0·4	Residual after roasting
3. *Minerals* (as oxide ash)	5·0	Figure slightly increased from green
4. *Acids* (in free state or as salts)		
chlorogenic	2·5	Residual (*c.* 40% rel.) after roasting
quinic	0·7	Largely developed on roasting
aliphatic	1·5	Largely developed on roasting; acetic, formic, malic, etc. (volatile/non-volatile)
5. *Carbohydrates*		
sucrose	0·0	Largely destroyed on roasting
reducing sugars	0·3	Balance of loss/generation
polysaccharides	32·0	Hydrolysable by acid to arabinose/galactose/mannose/glucose. Some destroyed on roasting, and incorporated in (10)
pectins	3·0	Contains minor sugars
6. *'Protein'*	10·0	Assessed by hydrolysis into amino acids. Some destroyed; some incorporated in (10)
7. *Amino acids* (free)	0·0	Reacted on roasting from green
8. *Coffee oil* (with terpenes, sterols, etc.)	18·0	Largely unchanged on roasting. Absolute figure increased slightly
9. *Lignins*	2·0	Unchanged
10. *Caramelised/condensation Products* (associated with other compounds; variously called 'humic acids', melanoidins, etc.)	23·2	Figure by difference, for carbohydrate-derived components and some chlorogenic acid and 'protein' breakdown components
Total	100·0	

Additional: volatile compounds (except acids) 0·1%; nicotinic acid 0·015%

2.4. Compositional Factors
2.4.1. Composition of Roast Coffee

The content of various components of roasted coffee has been fully discussed in Volume 1 under the different chapter headings for each group of compounds. Even so, there is no clear consensus from published information as to exact details of composition, which necessarily differs somewhat between the arabica and robusta types of coffee and particularly in the degree of roast to which the green coffee has been subjected. Differences will arise from natural variations within a species, and though ostensibly at the same degree of roast, from different roasting procedures; also some differences can be exaggerated by use of different analytical procedures. Roasted coffee also contains ill-defined pigmented/coloured/ high molecular weight compounds present in amounts very much dependent upon roasting degree, derived by caramelisation of the sugar (mainly sucrose) present and by condensation reactions of some of the polysaccharides originally present into which may also be incorporated some of the 'protein' and chlorogenic acid and its breakdown products. In practice, the composition of roasted coffee may be conveniently considered by the typical average content figures of an arabica coffee at a medium roast colour as given in Table 2, in which the caramelised/ condensation product content is given as a difference figure from 100% after the other components have been determined by direct analytical procedures.

The differences in composition for a medium roast robusta coffee from arabica lie chiefly in the amount of coffee oil, down by some 6%; caffeine content up by 1·0% or more; chlorogenic acid (residual and breakdown products) up by 1–2%. It is probable that the remaining constituents, particularly the polysaccharides and caramelised/condensation products, are up by some 5%, based upon the known difference of soluble solids extractable under the same conditions of about 3% higher for robusta (at 100°C) and of about 6% (at higher temperatures).

Differences in composition will occur for either arabica or robusta coffee at higher or lower roast levels. Though no reliable published information is available for each and every constituent, it is likely that the proportional content of caramelised/condensation products will markedly increase with darker degrees of roast. However, it should be noted that whereas a medium roast coffee implies a dry basis roasting loss of about 6%, that of a dark roast is much higher ($\approx 12\%$).

2.4.2. Yield and Composition of Extracts

The water-soluble dry solid content of roasted coffee as a percentage of roasted coffee (dry weight) is an important parameter in extraction. In instant coffee manufacture, this parameter is expressed as the so-called yield of soluble solids in the coffee extract, though it is calculable in a number of different ways each giving different numerical values. Yield for given conditions of extraction is the dry weight of soluble solids in the particular aqueous extract taken as a percentage of the roasted coffee weight used in that extraction (either a dry basis weight, or an 'as is' weight, which includes a known moisture content, say from 0–7%, in the roasted coffee). For an instant coffee, this yield may be calculated on the 'as is' weight of soluble solids (i.e. including its moisture content, say 0–5%). Furthermore, yield may be expressed not on the roasted coffee weight used, but on the original green coffee weight at a known moisture content (say 12%). This figure is referred to as the green yield in industrial practice, and is lower than the figure based upon roasted coffee, by the multiplying fractional factor of the roasting loss (from green to roasted coffee at the relevant moisture contents) which will be dependent upon degree of roast. Finally, a yield figure may be either the theoretically calculated figure or an actual figure which includes mechanical process losses of extract or dry product (or even from green to roasted). It is therefore clear that a figure given for 'yield' can be misleading unless it is stated clearly as to what it precisely refers (whether on roasted, green, dry, 'as is', and so on).

The soluble solids content of roasted coffee by extraction with hot water at 100°C has been discussed in Chapter 3, section 2.2. Here the figures for exhaustive extraction are of most interest. Household extractions range from 18 to 25% yield of dry soluble solids from dry medium roast coffee (see Chapter 8), and exhaustive extraction gives higher figures, approximately 28–30% (both sets for arabica coffee). Figures for robusta coffee are somewhat higher (2–3%) in each case. Approximate compositions of household extractions have been given by Vitzthum,[7] Maier[8] and others; all would agree that all the components in roasted coffee cited are extracted to a greater or lesser degree, except the coffee oil and lignin, which are left virtually unextracted from the grounds. The degree of extraction in household brews given by Maier[8] ranges from 100% for the quinic and aliphatic acids, 85–100% of the alkaloids and unchanged chlorogenic acids, 40–100% of the volatile compounds, 15–20% of the 'protein', and 20–25% of the melanoidins, down to 1·5% of the coffee oil. The unchanged polysaccharides comprise a range of substances of different molecular weight and structure, of which it is likely that any glucan (cellulose) will be quite insoluble at this or any other temperature unless in a very acidic medium. Full compositional data are not generally available for exhaustive extraction at 100°C, but it is likely that the somewhat higher yield found will be made up of a complete extraction of those substances like caffeine, and further extraction of others.

By extraction with water at temperatures above 100°C under pressure, say up to 180°C (of the grounds which have already been extracted at 100°C), it is known that there is progressive solubilisation of the compounds not already fully extracted; in particular, the 'protein', polysaccharides and melanoidins. There is therefore an increase in the percentage yield of soluble solids from the roasted coffee, with increasing temperature of extraction (all other conditions being the same), as illustrated in the data of Kroplien[9] presented in graphical form in Fig. 3. In this work, the extracts were taken stagewise from autoclave extractions for 1 h at each temperature; with extract separated off in successive extractions and fresh water added to the spent grounds for re-extraction. The yield figure at each temperature is therefore the cumulative figure. From these data there appears to be a linear increase of yield with temperature (not consistent with the Arrhenius equation), and the yield from robusta is markedly higher (c. 10%) than from two arabica coffees examined. Thaler[10] also prepared extracts at different yields by a technical method not disclosed, in which a medium roast Colombian arabica coffee showed up to 53% yield (dry roast coffee basis) and up to 58% for a medium roast Angolan robusta coffee. Kjaergaard and Andresen[11] reported a maximum yield figure of 60% for a roasted robusta coffee in their continuous countercurrent extractor.

Kroplien[9] used his extracts for analysis of their monosaccharide content; that is, of arabinose, galactose and mannose, as shown in Fig. 4 for the

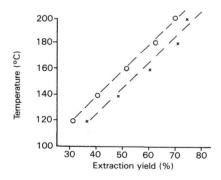

Fig. 3. Relationship between temperature (°C) in successive extraction stages with cumulative percentage extraction yield obtained (% of dry roasted coffee). ○--○, Colombian arabica; x--x, Zaire robusta roasted coffee. Data from Kroplien.[9]

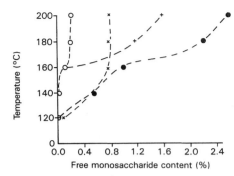

Fig. 4. Relationship between the free monosaccharide contents of arabica extracts from Fig. 3 and the temperature of extraction (°C). Contents based as a cumulative percentage of original roast coffee weight. ●--●, Total content; +--+, mannose; x--x, arabinose; ○--○, galactose. Data from Kroplien.[9]

Colombian arabica coffee. These contents can then be compared with corresponding contents of the monosaccharides and of polysaccharides (after complete laboratory hydrolysis to constituent monosaccharides) in the original roast and ground coffee. It is apparent that the actual amounts of these free monosaccharides are quite small, with the arabinose content progressively increasing with temperature of extraction but becoming constant at 160°C corresponding to a 52% yield of soluble solids (dry) of dry roasted coffee. On the other hand, the mannose content remains at first very low, but then increases markedly in amount. It is evident that these monosaccharides have been split off from their parent polysaccharides, which may give a clue to the way in which these high-molecular-weight substances are brought into solution at higher yield by a temperature-activated process. Some estimate of this splitting off, or 'hydrolysis', may be obtained by comparing these figures with the original contents of the polysaccharides in the roasted coffee; for example, contents given by Thaler and Arneth[12] arbitrarily stated as 'araban', 'galactan', 'mannan' and 'glucan' at 0·8%, 6·5%, 14·2% and 5·5% respectively (on a dry green basis) in a medium roast arabica coffee (though these may not be the true amounts, nor represent the actual polysaccharide make-up present, see Volume 1, Chapter 3). Calculation suggests that at 52% yield all the arabinose is split off, but only 2% of the galactose, 9% of the mannose and negligible glucose. The comparison made by Sivetz and Desrosier[13] with the industrial solubilisation by acid hydrolysis of cellulose seems to be a false analogy. Thaler and Arneth[12] themselves discount the marked significance of hydrolysis in coffee extraction.

Monosaccharide content may also be expressed as a percentage of the soluble solids in coffee extracts (rather than in the original roast and ground coffee). Such figures, also given by Kroplien, are very similar to those obtained directly by more modern HPLC techniques for various commercial instant coffees.[14,15] It is generally agreed that the amounts of glucose are very small even at the higher yields; the presence of fructose is indicative of chicory extract.

Thaler[10] used his extracts for analysis of their polysaccharide content from their monosaccharide content after laboratory hydrolysis; but also separately of the so-called 'high polymerics' (after release by use of chlorine dioxide, and subsequent dialysis). He was able to demonstrate a progressive extraction (expressed as a yield of carbohydrate from the original roast and ground coffee) starting from the lowest yields taken at 100°C extraction for the robusta coffee, though there was apparently a levelling-off for the arabica coffee, as illustrated in Fig. 5.

For the components that are already extracted at 100% from the roast and ground coffee at the lowest yield (say by exhaustive extraction at 100°C), it is evident that their content expressed as a percentage of the dry soluble solids will steadily fall with increasing yield taken, for example caffeine content and potassium/mineral matter content (as fully described in Volume 1, Chapter 2). Whereas, therefore, extracts taken at different yields from the same roasted coffee will contain essentially the same components at different yields taken, their relative proportions present will be different. Some changes will also occur if constituents fully

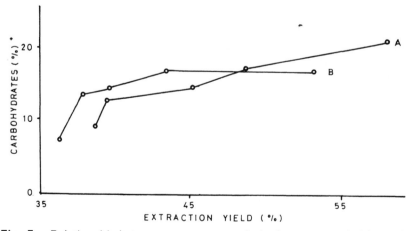

Fig. 5. Relationship between percentage carbohydrate content (with respect to roasted coffee weight) in extracts taken at different extraction yields; for (A) robusta, (B) arabica coffee. Reproduced from Thaler.[10]

extracted at a lower temperature are exposed for a time at a higher temperature, becoming, for example, insoluble. Such occurrences are largely avoided by stagewise procedures adopted in large-scale manufacture.

The acidity of coffee extracts and brews has been considered in Volume 1, Chapter 8. Kjaergaard and Andresen[11] report that their acidity as determined by pH measurements increases with yield (on roasted coffee taken) and residence time in continuous countercurrent extraction (and no doubt similarly for percolation batteries), though the differences are small, e.g. for a given blend and roast colour the pH is given as 5·1 at 30% yield and 4·8 at 50%. Such differences may not, however, be reflected in the final dried powder, and where volatile compound recovery is practised, contributions to acidity from formic and acetic acid may well be reduced. There are a number of patents for techniques to reduce acidity of instant coffee extracts as desired, e.g. use of chitosan as an absorbent, British Patent 2,029,888 (Nestlé, 1980); use of electrodialysis, European Patent 49-497B (Nestlé, 1985); use of ion exchange, European Patent 151-772A (Nestlé, 1985); and also after steaming roast and ground coffee, US Patent 4,303,686 (General Foods, 1981) and US Patent 3,420,674 (General Foods, 1969). There is also patent interest in reducing bitterness level in coffee, again as desired, e.g. British Patent 1,465,168 (Unilever, 1977); and Canadian Patent 940,920 (General Foods).

Changes in composition and flavour in coffee extracts can occur subsequent to extraction; that is on holding extracts at elevated temperatures (say above 60°C for several hours). Deterioration in coffee brews has been studied,[16,17] though there is no particular information published for high-concentration coffee extracts. Nevertheless, cooling of extracts is routine procedure. Volatile compound distillates of various kinds and coffee oil will be found to be highly unstable, requiring special care and low temperatures during holding. In the former it appears that, in the presence of water, interaction of components can readily occur,[13] leading even to precipitates of polymers, whereas with coffee oil, low oxygen headspace levels ($<4\%$) are especially required.

2.5. Balances and Rate and Productivity Factors
2.5.1. Mass and Heat Balances
In a percolation battery, the mass balance for the soluble solids may be conveniently set over a single cycle for an equivalent single column, as in Fig. 6.

In practice, the quantity of soluble solids (s) taken out in solution with the spent grounds will be very small in comparison with that (S) in the

ROAST COFFEE
IN

X kg including
x kg moisture

SPENT GROUNDS OUT
containing
s kg soluble solids at a yield of $y\%$,
w kg water and

$$X\left[1-\left(\frac{Y+y}{100}\right)\right] - x \text{ kg}$$

inert dry grounds

FRESH WATER
IN

= Feed water rate (F, kg min^{-1})
× Cycle time (T, min)

SOLUBLE SOLIDS OUT
[IN EXTRACT]

S kg solids at a conventionally
measured yield, $Y\%$

$Y = S/X \times 100\%$, W kg water
Therefore concentration
$= S/(S+W) \times 100\%$

Fig. 6. Generalised mass balance for a percolation battery extracting roast coffee. Diagrammatic, actual flows countercurrent with multiple columns.

extract. If the volume of a percolation column is known (say V litres), then from the dry weight of the roast coffee, the true inherent particle density of the spent grounds (say 1·20) and the density of the water at the discharge temperature (say 0·844 kg per litre at 175°C) it is possible to calculate the weight of water discharged from the column with the spent grounds; thus:

$$w = 0.844\left[V - \left(\frac{X-x}{1\cdot 20}\right)\right] \text{kg}$$

This amount of water will, however, decrease in amount on discharge to atmosphere pressure and a temperature of 100°C, by a figure which can be calculated by enthalpy difference (e.g. a 13% decrease).

Similarly, the weight of water associated with the soluble solids in the extract is given by

$$W = (FT + x - w) \text{ kg}$$

It is then possible to determine the expected solubles concentration for a given yield (Y) obtained, the ratio of FT/X (i.e. the effective water to coffee ratio) for any cycle time, the column volume/weight of coffee ratio, and the moisture content of the roast coffee ($m = x/X$). If, for example, the water to coffee ratio is 3:1, the yield on roasted coffee is 45%; and if the column volume to coffee weight ratio (V/X) is 2·63, then it follows that the concentration of solubles in the extract will be 22·5%, irrespective of the absolute weights of the coffee or volume of the column (i.e. V/X

fixed, specific volume of fill weight). In operating practice, a ratio known as the extraction or draw-off factor is commonly used (weight of extract drawn/weight of roast coffee in column, i.e. $W + S/X$). In this example, its value is 2·00. From this mass balance also, the percentage of water in the final grounds is given by

$$\frac{w'}{w' + X\left[1 - \left(\frac{Y+y}{100}\right)\right]} \times 100\%$$

where w' is the corrected value of w after discharge to atmospheric conditions; $w = FT + x - W$.

Though a typical water to coffee ratio might be 3:1, according to the number of columns in the extraction battery, each unit weight of roasted coffee will actually be exposed to [$(FT/X) \times$ number of columns on stream] weight of water in its total residence or contact time in the battery.

In a mass balance for a continuous countercurrent extractor, a very similar diagram can be drawn, except that the balance should be on an hourly basis (or contact time basis). In the percolation battery, the concentration of soluble solids is that of the whole of the extract batch after drawing, which will gradually decrease from the start of drawing to its finish. In the continuous extractor, the concentration is the same at any time during extraction, and the term 'draw-off' has no meaning. The effective volume of the extractor for the time basis also has to be used. A mass balance could be similarly constructed for the volatile compounds, either singly or collectively.

Heat balances can also be constructed, taking a datum level say at the ambient temperature of the roasted coffee, and assuming the inclusion of all heat inputs and outputs. Such balances, somewhat difficult to attain accurately in percolation batteries, have no particular practical value, except to indicate the heat losses and therefore insulation that might be required to maintain a particular temperature profile.

2.5.2. Mass Transfer Rates

2.5.2.1. General considerations. The mass balances just described of course give no indication of the yield or efficiency of extraction of soluble solids or volatile components obtainable under given operating conditions; mass transfer rates need to be known. There are a fair number of independent parameters (also many dependent ones) which affect the kinetics of transfer; e.g. for percolation batteries we have feed water temperature and temperature profile, feed water rate (which will determine liquid velocity past the solid particles, cycle time, and contact time according to the number of columns on stream), grind size of particles (average and range) and draw-off factor (as previously defined); all for a given blend of coffee and geometry of construction of the columns, which if altered will introduce further parameters. Similarly, for continuous extractors, feed water temperature and profile, feed water rate (determining contact or residence time), roast coffee feed rate, screw speed and grind size of the coffee must be considered.

Various empirical relationships will be found to be valid for particular percolation batteries; though in general it will be found that increasing the draw-off factor (all other conditions being the same, or nearly the same) will increase the yield of solubles taken, as will increasing the feed water temperature (say up to 180°C). The effect of other variables has to be carefully considered for particular cases, e.g. choice of grind size (see section 2.5.3). The initial wetting of the roast and ground coffee is an important element in extraction, and occurs in the fresh coffee stage as the coffee is being contacted with liquid for the first time. It is necessary that there be a sufficiently high liquid temperature, i.e. between 70 and 100°C, to facilitate this wetting, and that extraction be allowed to proceed thereafter as quickly as possible.

For continuous extractors some published data have been provided by Kjaergaard and Andresen[11] on their performance under different conditions. For example, Fig. 7 shows that the yield of soluble solids obtainable from an unspecified arabica–robusta blend approaches asymptotically a stated maximum of 57% (based on dry roasted and ground coffee) at a capacity rate markedly influenced by the applied temperature profile. A residence time of 90 min would be typical in a 2·7 m^3 (2700 litres) volume extractor for roasted coffee to give a yield of about 45% dry basis (or 79% of the assumed total soluble solids), when a temperature profile of 100–175–175°C is used. Water to coffee ratios and therefore concentration of extract are not given, though the former will be of the order of 3:1.

Even less has been published on the kinetics of removal of volatile compounds; though in home brewing, with its single-stage extraction, Pictet and Vuatez[18] have indicated the importance of high water to coffee ratios in achieving high efficiency of solubilisation. British Patent 1,547,242 (Nestlé, 1979) states that greater draw-offs or, in effect, more-dilute extracts yield a more desirable beverage flavour and aroma. Similarly, British Patent 2,005,126B (Douwe Egberts, 1983) comments on the role

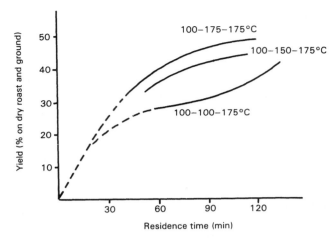

Fig. 7. Relationship between percentage yield of soluble solids (based upon dry roast coffee) and residence time in a Niro continuous countercurrent extractor at different temperature profiles. Courtesy of Niro Atomizer A/S, Copenhagen.

of the water to coffee ratio in securing high recovery of volatile compounds. It is evident that as much as possible of the volatile compounds should be extracted before the coffee is contacted with water of increasing temperature much beyond 100°C, and that industrial multi-stage extraction has a potential for extracting more and more efficiently these important compounds.

2.5.2.2. Rate equations. Mass transfer in the extraction of solids by a solvent has been studied in its fundamental aspects by various authors, notably Schwartzberg[2] of the University of Massachusetts; Spanninks[20] and Bruin and Spanninks[19] at Wageningen, The Netherlands; Besson[21] of Nestlé; and Loncin[22] of the University of Karlsruhe. Its full mathematical study presents numerous problems in establishing satisfactory models, especially if further applied to the complete range of roast coffee extraction conditions for instant coffee (e.g. differential rates for different components, and non-isothermal and non-physical factors).

The mathematical approaches taken by each of these authors differ. At the heart of the matter is the relative importance of two separate potential rate-determining factors. First, there is the resistance to mass transfer resulting from 'intra-particle diffusivity', that is, the solute has to diffuse through the cellular matrix of the solid particles to the surface, generally according to Fick's laws of diffusion. Secondly, there is the conventional liquid film resistance at the surface of the extracted particles with the bulk extracting liquid, dependent upon the external liquid flow conditions. The relative significance of these two resistances is governed by the dimensionless Biot number, $Bi = k_1 . r/D$, where k_1 is the mass transfer coefficient for the liquid film, D is an effective diffusivity of the solute within the solid phase and r is a distance dimension equal to the radius in a spherical particle. A higher value of D means lower resistance to internal diffusion. The higher the value of Bi, the greater the significance of intra-particle diffusion; and the lower, the greater the significance of the liquid film (i.e. intra-particle diffusivity can then be neglected).

Charm,[23] for example, considers the extraction of roasted coffee in terms of a liquid film resistance only, so that

$$\frac{dW}{dt} = k_1(C^* - C)$$

where W is the amount of solute transferred in time t across a liquid film of resistance (k_1) with C and C^* the actual and equilibrium concentrations. If V is the volume of the extractor, then $dW = VdC$, and if A is the surface area of the particles, then on integration, from initial to final conditions,

$$\int_{C_{in}}^{C} \frac{dC}{C^* - C} = \int_0^t \frac{k_1 A \, dt}{V}$$

or

$$\ln \frac{C^* - C_{in}}{C^* - C} = \frac{k_1 A t}{V}$$

When $C_{in} = 0$, i.e. no solute in feed liquid (water), this equation simplifies to

$$1 - \frac{C}{C^*} = e^{-k_1 A t/V}$$

A sample calculation is provided in which the extraction yield of coffee solubles is taken but only at 24% of the roasted coffee.

Schwartzberg[2] and Desai and Schwartzberg[24] consider intra-particle diffusivity to be paramount in roast coffee and sugar beet extraction, and provide equations on this basis alone for countercurrent extraction to obtain yields and solute concentrations in extracts. The solutions to such equations[25] (see also Chapter 6 for volatile retention in spray drying, and Chapter 7 for diffusion of gases from roasted coffee) also have an exponential character but incorporate the dimensionless Fourier number,

$Fo = Dt/r^2$, where D is a diffusion coefficient for the solute in the solid particles, t is the residence time in the extractor, and r is a distance dimension equal to the radius for a spherical (or infinite cylindrical) particle. Schwartzberg[2] indicates, for example, a leaching yield of 96% (of the available sucrose) from sugar beet cosettes (3 mm thickness) with a D value for the sucrose of $1 \times 10^{-9}\,m^2\,s^{-1}$ for a solids hold-up time of 64 min in a continuous countercurrent system. In discussing the extraction of roasted coffee in a diffusion battery, he states that the diffusivity of coffee soluble solids in grounds is only about one-tenth of that of sucrose in beets at 75°C; though with higher temperatures operating, the average diffusivity will approach that of sucrose.

Spanninks[20] explains in detail two possible different approaches in using mass transfer equations for extraction. The first method revolves round the intra-particle diffusion equation (with an effective diffusion coefficient), together with the liquid film mass transfer coefficient, to give a solution in terms of the Fourier and Biot numbers, and generally requires the use of a computer. The second method, which is initially mathematically simpler, centres on the use of an overall mass transfer coefficient, in a so-called 'lumped resistance model' involving assessment of transfer units. As in the well known two-film theory, this overall transfer coefficient is made up of the liquid film resistance (readily determined) and the 'diffusion resistance' (not strictly a film), which is more difficult to determine. The application of the second method is described in detail by Bruin and Spanninks[19] for a diffusion battery. The following is a guide, including their use of symbols. By the number of transfer units concept, applicable to extraction in general, the separation or extraction obtained is indicated by the number of 'exterior apparent' units given by

$$N_{ext} = \int_{C'_{in}}^{C'_{out}} \frac{dc'}{c'* - c'}$$

where c'_{in} and c'_{out} are the inlet and outlet concentrations of the solute in the continuous liquid phase ($kg\,m^{-3}$ units), and $c'*$ is a hypothetical local concentration which is the concentration of solute that would be in equilibrium with the solid phase if the phases were contacted purely in countercurrent plug flow. This equation is solvable in terms of two other parameters, an extraction factor Λ and the required separation efficiency η, by the Colburn equation:

$$N_{ext} = \frac{1}{1-\Lambda} \ln\left(\frac{1-f\Lambda}{1-f}\right)$$

where η is defined by the ratio of the actual amount of solubles transferred and the amount transferred when the solids leaving the extractor are in equilibrium with the liquid inlet concentration, at flow rates ϕ_E(extract) and ϕ_R(solids):

$$\eta = \frac{\phi_E(c'_{out} - c'_{in})}{\phi_R(\omega'_{in} - c'_{in}/m)} = \Lambda \frac{c'_{out} - c'_{in}}{m\omega_{in} - c'_{in}}$$

where ω'_{in} and ω'_{out} are concentrations of solute in the solid phase; m is the slope of the equilibrium line dc^*/dw'; f is merely η/Λ. A graphical plot of N_{ext} vs. Λ for different values of η is available.[19] In many cases, where $c'_{in} = 0$, the equations will be simpler.

Only when there is true countercurrent plug flow does the value of N_{ext} equal the true number of transfer units as given by

$$N_{t,c} = k_{0c} \cdot a \cdot V/\phi_E$$

on an overall liquid phase basis; where a is the specific interfacial area per unit of packed volume (m^2 m^{-3} units); V is the volume of the extractor; and k_{0c} the overall mass transfer coefficient. In general, $N_{t,c}$ will exceed N_{ext} by a factor which takes into account deviations from true countercurrent flow due to cross-flow, axial dispersion and so on. This factor has to be determined for particular cases, and with an estimate of the overall mass transfer coefficient, residual unknowns (e.g. liquid residence time, $t = \phi_E/V$) can be calculated.

In the particular case of a fixed bed as in Fig. 8, the diffusion of the

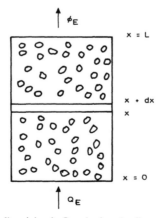

Fig. 8. Diffusion in a fixed bed. Symbols: ϕ_E, flow rate of extracting fluid; x, distance variable with L, actual length of bed; A, cross-sectional area of extractor, m^2; and h, interstitial void fraction (or liquid hold-up).

solute to the liquid is in a non-steady state and a macroscopic mass balance is not valid.

The local mass transfer rate of solute from solid to liquid has to be determined, and is given by the mass balance. In the liquid phase,

$$ALh \frac{\partial c'}{\partial t} = k_{0c} aAL(c'^* - c') - \phi_E \frac{\partial c'}{\partial x/L}$$

and in the solid phase,

$$AL(1-h) \frac{\partial \omega'}{\partial t} = -k_{0c} aAL(c'^* - c')$$

There are boundary conditions, i.e. at t (time) = 0, $\omega = \omega_0(x)$; $c' = c'_0(x)$; at $x = 0$, $c' = c'_{in}(t)$.

These equations can be made dimensionless for the liquid and solid phase concentrations, distance and time, and include the number of *true* transfer units on an overall liquid phase basis. Spanninks presents solutions to obtain, say, the emergent concentration, c'_{out} after a time t. Such an analysis would apply to fixed bed extraction, as occurs in household coffee brewing by filter methods; but also to the first stage of a battery or when physical solubilisation of coffee solubles is taking place.

With a diffusion battery, consisting of a number of fixed beds or columns exceeding about four, the battery approaches a truly countercurrent system (i.e. $N_{ext} \to N_{t,c}$). Spanninks and Bruin provide a relationship from which $N_{t,c}$ can be determined from N_{ext} for a given number of columns, and a modified value of Λ (including the cycle time); and allows for the fact that the liquid filling the interstitial voids in the first column is discharged with the exhausted solids.

A similar mathematical analysis is not, however, given for the continuous countercurrent type of extractor, in which the movement of solids is provided by means of a screw, which will cause cross-flow as well as countercurrent; for this the particular relationship between $N_{t,c}$ and N_{ext} would be needed.

In order to use these equations fully (e.g. to calculate residence times, extractor volumes, etc., for either design or performance analysis), it is necessary to obtain a value of k_{0c} (or $k_{0,d}$, on a solids (dispersed) phase basis), made up of the liquid (continuous phase) film resistance, k_c, and the particle diffusion resistance, k_d, where[3]

$$\frac{1}{k_{0,d}} = \frac{1}{mk_c} + \frac{1}{k_d}$$

Correlations in terms of Sherwood, Schmidt and Reynolds numbers for the flowing liquid to determine k_c are well known.

Reynolds numbers include the velocity of the flowing liquid (of known significance in the rate of coffee extraction), suggesting that the liquid film resistance is also of importance and may be controlling under certain conditions. A specific method is explained to obtain a value of k_w (which varies throughout the extraction), as a so-called asymptotic value. This mass transfer coefficient includes the solute diffusion coefficient required to express the effect of intra-particle diffusion.

Besson[21] takes the first approach in analysing the variables and presenting equations, i.e. solving a continuity equation with the intra-particle diffusion equation with boundary conditions that include the liquid mass transfer equation at the surface. He also considers a possible skin effect at the solid particle surface, as suggested by Loncin.[22] Three types of extraction are considered: a well stirred batch vessel, a fixed bed column and a continuous countercurrent extractor; and both intra-particle diffusivity and liquid film mass transfer are considered. Solutions are presented in terms of Fourier, Biot and other dimensionless numbers (liquid to solid ratio and normalised porosity), in particular for a required extraction efficiency (i.e. percentage of original solute present extracted) rather than concentration–time curves. With a fixed bed column he presents graphical plots of Fourier number against the liquid to solid ratio at a fixed extraction yield of 90% (of the available solute, i.e. not of yield from the total solids) when in fact the Biot number is varying, which shows the effect of column length. He concludes that providing the Biot number is less than 50, column geometry will play an important role in the leaching kinetics (influence of surface film coefficient). In scale-up, the conditions on the same curve on this plot (involving a length parameter) have to be observed; whereas with a longer column the same extraction yield will be obtained with reduced extraction time and liquid flow.

For complete application of these equations in specific cases of extraction, knowledge of the equilibrium relationship of the solute concentration between that in the solid phase and that in the liquid is required, i.e. the value of m in the foregoing discussion. Such information is often presented in the form of a ternary equilibrium diagram (inert solids–solute–solvent). For coffee solubles extracted at 100°C between grounds and water, information is available, as discussed by Sivetz and Desrosier,[13] though a value of m is not given but can be obtained from simple laboratory work. However, the coffee solubles in industrial extraction in contact with fresh coffee at around 100°C also come partly from extraction at higher temperatures, which may influence this value of m.

Over the whole system of an industrial coffee extraction, application of these equations is further complicated by differing mass transfer rate factors at differing temperatures from inlet to outlet, and furthermore rate factors may have a contribution from rates determining initial solubilisation by non-physical means, e.g. release from covalent and other linkages of polysaccharide substances with cell walls, etc., resulting in the formation of some monosaccharides as already noted by Kroplien.[9] It is not surprising, therefore, that considerable empiricism at present is required to unravel the effect of various parameters in optimising extraction conditions. Nevertheless, these equations have a role in the correct understanding of the mechanisms involved, and in guidance on practice.

2.5.3. Productivity

In commercial extraction, productivity rates are an important directly observable factor, that is, kilograms of soluble coffee solids produced per hour, at the required yield and at an economic concentration of the soluble solids in the extract. The capacity for an extractor, a less useful figure, is based on the throughput of the roast and ground coffee per hour. Optimisation of the operation of a given extractor will be sought. Increasing yields, however, can generally only be obtained at lower productivity rates, and with lower concentrations of soluble solids.

Some of these factors can be seen from the data for a single-unit continuous countercurrent screw extractor.[11] Figure 7 has shown that as the yield (dry solubles on dry roasted coffee) reaches a maximum, so the required residence time (of the solids) determined by the screw speed increases exponentially (and the capacity similarly decreases). This exponential effect is to be expected from the considerations of sections 2.5.1 and 2.5.2. A further diagram (Fig. 9) from the original paper indicates that for this extractor (of $2.7\,m^3$ volume) for the same temperature profile, the yield obtainable at $400\,kg\,h^{-1}$ roast and ground coffee capacity is 42% and therefore a productivity rate of $168\,kg\,h^{-1}$ dry solubles; but at $600\,kg\,h^{-1}$ the yield is only 30% at $180\,kg\,h^{-1}$ dry solubles productivity. It is not clear from the paper whether the same water to coffee ratio was used, as changes in this ratio will also affect the important concentration of soluble solids in the extract and the actual yield obtainable.

With percolation batteries, the key factor in capacity is the cycle time

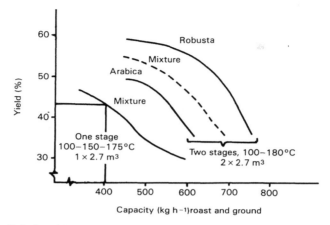

Fig. 9. Relationship between percentage yield of soluble solids and capacity throughput of roast and ground coffee in a Niro continuous countercurrent extractor for the same mixture (blend) as in Fig. 7. (Diagram also shows relationships in a two-unit extractor.) Courtesy of Niro Atomizer A/S, Copenhagen.

for the extraction that is achieved. The lower this figure, the greater will be the capacity in terms of roast and ground coffee input per hour, and thus, provided the yield is not greatly changed, the greater will be the productivity in terms of soluble solids. With a given number of columns on stream, decrease of cycle time means higher velocities of liquor through the coffee bed, which in turn can mean escalating pressure drops through the columns, beyond the capability of the system. High-pressure drops in beds involve a number of physical phenomena, and are likely with columns of large volume (and also of large diameter) filled with relatively fine and non-uniform grind coffee, and from which high yields are taken. The voidage for passage of liquor becomes reduced through the migration and segregation of fines and through the compressibility of the coffee bed (a function of the condition of the coffee and of the pressure applied). Increasing the number of columns on stream for a lower cycle time within the same overall residence time may also be ineffective on account of the longer travel path. The incidence of excess pressure drops is capable of fundamental analysis, though little has been published. For the same reason, efforts to increase yield for the same output can be frustrated. In the continuous countercurrent extractor, however, with the screw action on the grounds, liquor flow is not so impeded.

2.5.4. Water

The feed water used for percolation is normally taken direct from the local mains water supply, but when this water is 'hard' with calcium and magnesium salts, it is usual to soften it before use by conventional methods. Softening is practised not so much for reasons of the coffee extraction, but rather to prevent scaling in the feed water-heating and other equipment. Softening will introduce sodium into the coffee extract, e.g. feed water at 400 ppm equivalent calcium carbonate will give some 184 ppm sodium on softening, and thus between 700–900 ppm in the resultant instant coffee (dry solubles basis), depending upon concentration of the extract from percolation. The feed water may of course be demineralised after softening if considered necessary.

Studies have been carried out in home-brewing methods of extraction with mineral-containing waters with some observable effects, e.g. longer time of filtration with hard water, and increase in pH value with softened water.[17] However, no such effects appear to have been established in industrial coffee extraction. The use of a small amount of dilute mineral acids would slightly increase the yield of coffee soluble solids in extract, and increase speed of extraction; this acidity could then be corrected to normal by addition of alkali. This practice is not, however, widespread (flavour may be affected), much greater preference being given to the simple use of potable water. The EEC Coffee Directive of 1977, indeed now adopted, does not allow the use of either acids or alkalis in coffee extraction. There are, however, a number of patents for the use of dilute acids in coffee extraction at 100°C, e.g. US Patent 2,687,355 (National Research Corporation, Cambridge, Mass., 1954).

The reuse of water from the spent end of a percolation battery has also been proposed, with a purification process before re-entry into the system, e.g. US Patent 3,972,809 (Nestlé, 1976).

2.6. Volatile Compound Handling

2.6.1. Prestripping from Roast and Ground Coffee

There are numerous patents covering the prior removal of volatile compounds from roast and ground coffee before its conventional extraction with water, which then becomes essentially an extraction for the soluble coffee solids only. Steam under various conditions is the favoured vehicle, which can then be condensed to a mixture of liquid water and the volatile substances (generally miscible with the water, though this depends upon the particular volatile compounds and their concentration in the water). A selection of these patents, in which this technique is

Table 3
Some patents for the steam stripping of roast and ground coffees

Mode of operation	Patent/Date	Assignee
Steam condensates for admixture with coffee oil	BP 982,251 (1965)	Nestlé
Steam volatile flavour	USP 3,132,947 (1964) BP 1,206,297 (1967)	General Foods Corporation
Downflow steaming	USP 3,653,910 (1974)	General Foods Corporation
Steam condensates	BP 1,424,263 (1976)	Procter & Gamble
Continuous steaming	USP 3,244,582 (1966)	General Foods Corporation
Continuous steaming	Swiss P 563,792 (1975)	Nestlé
Continuous steaming	BP 1,446,881 (1976)	General Foods Ltd (UK)
Steam stripping	USP 4,204,464 (1980)	Procter & Gamble

practised or forms the main object of the claims, is given in Table 3. Most of these patents refer to steaming of the roast coffee already held within a percolation column of a battery, which is a convenient technique, since, after steaming, this column can be placed on stream in the conventional manner. The last three, however, propose the use of a separate item of equipment, so that the roast coffee after steaming, which will be damp though generally free-flowing, has to be mechanically fed into the fresh column of a percolation battery or into a continuous extractor. A variety of steaming conditions including condensation are proposed; though it is self-evident that the initial steam pressure has to be sufficiently high to pass the steam through a static bed of coffee in a column in a reasonable time. Although superheated steam could be used, it is generally preferable for temperature reasons to allow condensation–distillation at an advancing front to take place.

A fundamental analysis of this operation has not been published, nor do the techniques appear to have been applied to other foodstuffs; the efficiency of recovery will relate to the relative volatilities of the individual flavour compounds (see section 2.6.2). These condensates can be incorporated in the finished product in a number of different ways (see section 2.6.5).

An alternative to steam stripping, the use of inert gases such as carbon dioxide as the carrying agent, has been proposed, e.g. British Patent

1,106,468 (Nestlé, 1968) equivalent to US Patent 3,532,506. Supercritical carbon dioxide (dry) has also been proposed, e.g. German Patent 2,119,678 (Hag, 1971), though this solvent will also extract a proportion of the coffee oil. There are also a number of patents describing the use of various organic solvents, but these will not be further discussed as instant coffee by various legislations (e.g. EEC Coffee Directive, 1977) is only to be obtained by the extraction of roast coffee with water.

2.6.2. Stripping of Coffee Extracts

There are also numerous patents covering the removal of volatile compounds present in coffee extracts, prepared by extraction in the conventional or an alternative manner prior to drying. A selection of such patents is given in Table 4. The extracts are partially evaporated and the vapours condensed to give a mixture of liquid water and volatile compounds. Similar procedures and patents exist for the evaporative or flash stripping of fruit juices. The patents in Table 4 differ in operating procedures and conditions of stripping, and the last refers to the stripping of condensates already obtained by evaporative stripping of coffee extracts.

Extracts will preferably be continuously stripped as they are produced, giving advantage to the continuous systems of extraction. Again, a variety of evaporative stripping conditions have been proposed, which centre essentially around the pressure (or rather vacuum) and therefore temperature of evaporation and upon the percentage of water in the extract that should be evaporated to achieve the required removal of the desired volatile compounds. Thijssen and his colleagues have especially examined this operation in its fundamental aspects.[26,27] In continuous operation,

Table 4
Some patents for the stripping of coffee extracts

Patent No.	Date	Assignee
USP 3,132,497	1964	General Foods Corporation
USP 3,244,530	1966	General Foods Corporation
USP 4,092,436	1973	Procter & Gamble
BP 1,265,206	1972	Procter & Gamble
BP 1,547,242	1979	Nestlé
BP 1,563,230	1980	General Foods Ltd (UK)
BP 2,005,126B	1983	Douwe Egberts
EP 56-174B	1985	General Foods Inc. (Canada)

they showed that the following equation is applicable for equilibrium steady-state stripping:

% Stripping efficiency for volatile compound $(j) = \dfrac{\alpha_j}{(\alpha_j - 1) + F/V} \times 100$

where the stripping efficiency represents the percentage of the volatile component in the vapour over that in the feed extract; α_j is the relative volatility (related to water at 1) of the component j and F/V is the ratio of water in the feed extract to that of the vapour in a given time. The relative volatility of compounds has a known value, especially as the relative volatility at infinite dilution (α_j^∞), which is generally applicable to coffee extracts. Values can range from less than 1 (for acetic acid), to very high values (e.g. acetaldehyde at 120, and even much higher, as for higher aldehydic homologues). Values are temperature dependent, though generally not markedly so up to 100°C; and although their values in water are readily obtainable, there can be some effect (increase in value) as a result of the presence of dissolved substances such as sucrose or coffee solubles. Application of the equation by Thijssen[27] shows that by evaporating 80% of the water virtually all the acetaldehyde, for example, would be removed, whereas for acetic acid 31% would still be retained. In practice, in fruit-juice stripping, and in the examples given in the coffee patents cited, the amount of evaporation recommended for recovery of volatile compounds ranges from a few per cent to about 70%. A similar relationship can be derived (first by Rayleigh in 1888) for batch non-steady-state stripping, while with certain kinds of evaporators, such as the Centritherm with very fast evaporation, there is non-equilibrium evaporation leading to greater retention of volatile compounds. Stripping of extracts can also be carried out with steam in a countercurrent packed column.

Some stripping takes place as the extracting liquor moves up to a fresh stage of a percolation battery, or along the initial length of a continuous extractor. The liquor displaces entrapped air/gases (carbon dioxide) from the roast coffee, taking with them a proportion of volatile compounds which can be lost if not subsequently condensed, as discussed by Sivetz and Desrosier[13] as so-called vent gases. There are a number of patents for collection procedures.

2.6.3. Condensation of Volatile Compound—Steam Vapours

The vapour mixtures described under section 2.6.1 and 2.6.2 have to be condensed as effectively as possible, or to the extent required. Although very cold refrigerants (e.g. liquid ammonia and dry ice) are clearly effective, sufficient condensation of many compounds can be practised with the use of chilled/cooling water flowing in a tubular condenser. In any condensing system, the non-condensible gases (which arise from entrapped gases, e.g. in the roasted coffee or extracts; and by leakage of air into stripping equipment, especially when operating under vacuum) will carry volatile compounds with them and are inevitably lost, in an amount also dependent upon temperature and pressure in the condensing system. The amount of these gases should be kept as low as possible. This subject is discussed by Clarke and Wragg[28] who provide a working formula for various conditions, primarily related to the ratio of the volume of gases (A, m^3 min^{-1}) to the liquid condensate rate (B, g min^{-1}) and the partition coefficient (k_j) of dry volatile component (j) between air and liquid water. The latter figure corresponds to the relative volatility (usually taken at infinite dilution). The loss of volatiles is given by

$$\frac{7 \cdot 35 \times 10^2 \times k_j A/B}{(7 \cdot 35 \times 10^2 \times k_j A/B) + 1} \times 100\%$$

The value of k_j for acetone, for example, is 8·8 at 25°C and a pressure of 203 mm Hg absolute. This paper also describes an economic method for satisfactory condensation, British Patent 2,086,743B (General Foods Ltd, 1983).

Two-stage condensation is also recommended in a number of patents (e.g. British Patent 982,521 already cited) in that a first condensation at around 90°C (containing water and only a few volatile compounds) is discarded, whereas that condensed (about 5–20% of original vapour) at 20°C is retained. Condensates, dependent upon the original degree of evaporation in evaporative stripping, clearly can contain large amounts of water, so that fractionating distillation columns are often recommended for removing a substantial proportion of this water (which will also remove some of the compounds with low relative volatility, close to water) as 'bottoms', with a concentrate as 'tops'. Stripping the condensates has also been recommended (e.g. European Patent 56-174B already cited).

2.6.4. Coffee Oil Expression

As already mentioned, coffee oil present in roasted coffee (up to about 18% w/w in arabica coffee, see Volume 1, Chapter 5) contains a large proportion of volatile aroma/flavour compounds, which can be expressed from roast coffee by mechanical means (e.g. screw presses), as described

Table 5
Some patents for the purification of coffee oil

Patents No.	Date	Assignee
USP 3,704,132	1972	Procter & Gamble
BP 1,339,106	1972	
USP 4,156,031	1979	General Foods Corporation
German P 2,630,580	1985	
EP 43-467A	1985	Nestlé

by Sivetz and Desrosier.[13] A coffee oil, though bereft of its volatile components, remains in the spent grounds from aqueous extraction, so that spent grounds coffee oil may be obtained similarly by expression methods; indeed, some may also be present in the discharge liquor from the spent end of an extraction system. These latter oils are readily purified (from oxidised products, etc.) by conventional methods; additionally, the kahweol/cafestol of the unsaponifiable portion can be removed. A selection of such patents is given in Table 5.

2.6.5. Reincorporation of Volatile Compounds

Volatile compounds of roast coffee can be approximately divided into those that give rise essentially to so-called headspace aroma from the dry product, and those that essentially provide flavour to a prepared brew (together with headspace aroma from the brew). The former, present in very small amounts, are highly volatile and might readily be lost in subsequent drying operations. Consequently, there are numerous patented processes in which coffee oil (either original or fortified with aromatics from other sources) is 'plated-on' in small quantities (of the order of 1·0% or less) to the final instant coffee, usually simultaneously with packing into jars or other containers. A selection of patents is shown in Table 6.

Table 6
Some patents for plating coffee oils into instant coffee

Mode of operation	Patent No., date	Assignee
Spray plating	USP 3,148,070 (1964)	Afico AS (Nestlé)
Admixture of steam condensates	BP 982,521 (1965)	Nestlé
Coffee oil as an emulsion foam	BP 1,525,808 (1978)	Nestlé
	USP 4,072,761 (1978)	
Aroma-rich oil	USP 3,769,032 (1973)	Procter & Gamble

Table 7
Some patents for preparation and use of aroma frosts

Patent No.	Date	Assignee
USP 3,021,218	1962	General Foods Corporation
USP 3,821,447	1974	General Foods Corporation
USP 3,769,032	1973	Procter & Gamble
USP 4,007, 291	1977	General Foods Corporation

The coffee oil may be fortified with the steam distillates already described, or from so-called 'grinder gas', i.e. the gases carrying volatile compounds, particularly released during grinding and trapped as an 'aroma frost', for which there are a large number of patents (Table 7).

Two further patented methods for supplying headspace aromas may be mentioned: cryogenic transfer from roast and ground coffee on to instant coffee held at low temperatures (US Patent 3,823,241, Procter & Gamble, 1974), but also from heated roast coffee (European Patent 144-785A, Nestlé, 1985); and addition of so-called micro-porous roast coffee particles (British Patent 2,074,007B, General Foods Corporation, 1984, and US Patent 4,313,265 (1982)).

Although incorporation of suitable volatile components in a plating oil may confer flavour advantage also, another recommended procedure (see patents of Tables 3 and 4) is to take the volatile flavour concentrates, whether from steaming roast and ground coffee or from coffee extract evaporation, and add them to coffee extracts at a higher level of concentration of soluble solids (than achievable by extraction alone) from evaporative or other water-removal processes. These are then dried as described in Chapter 6. Direct coffee oil addition to coffee extracts followed by homogenisation has also been proposed, e.g. US Patent 3,244,533 (General Foods Corporation, 1966). The preparation of specially dried coffee flavour concentrates and addition to conventionally spray-dried coffee powders has also been proposed, e.g. British Patent, 1,541,895 (General Foods Ltd, 1976).

A further method of incorporation of volatile components that has been proposed is in the form of small capsules, e.g. British Patent 2,028,093A (Nestlé, 1980) in which hardened capsules are prepared from 1 part coffee oil and 1 part roast coffee distillates, which are then admixed with instant coffee granules. The interest of such capsules is in providing the so-called brew aroma, the slow release of headspace aroma from a prepared cup of coffee. This release can be made more effective by having such capsules float to the surface, and this can be assisted by foamed capsules, e.g. US Patent 4,520,033A (Nestlé, 1985).

It has also been proposed to add volatile flavour compounds of coffee flavour significance that are not directly derived from though present in coffee (or nature-identical), e.g. addition of methyl mercaptan to extracts before spray-drying, in US Patent 3,540,899 (General Foods Corporation, 1970). There has also been a considerable study of sulphur volatile compounds as reaction products of carbonyl compounds, hydrogen sulphide and/or ammonia, and their proposed use in coffee extracts (US Patents 3,676,156 (1972) and 3,873,746 (1975)). Firmenich et Cie of Switzerland have patented the use of a whole range of prepared volatile compounds suitable for coffee flavour modification or enhancement (e.g. US Patent 3,922,366 (1975); and many others, as in British Patents 1,156,472–490). In the context of 100% or pure coffee extracts, such additions are not practised.

3. PROCESS EQUIPMENT

It is probably fair to say that most of the equipment used for carrying out the processes described in this chapter will be designed by the major instant coffee manufacturers themselves; though specific items of plant in an ancillary role, but also whole installations, are available from a number of plant manufacturers. Sivetz and Desrosier[13] and the earlier volume[29] set out to describe details of percolation batteries and ancillaries together with practical hints on operation including start-up, some of which are inevitably controversial. There have been numerous articles on coffee extraction of a trade journal character, especially in the 1950s and 1960s, though these are now outdated.

3.1. Percolation Batteries

The general principles of their basic operation have already been given in section 2.1. In particular, Niro Atomizer A/S of Copenhagen, Denmark, supply complete percolation battery plants, generally with columns of tall/narrow dimensions. A complete large plant needs mobile containers on overhead conveyors for holding and then loading into the columns the required weighed charges of roast and ground coffee. The columns themselves require well designed, reliable and pressure-tight valves both at the top and, especially, at the bottom through which the spent grounds have eventually to be discharged. The spent grounds have to be disposed of, and a typical system would involve separation of the free liquor and pressing of the grounds by screw presses to about 60% moisture content. Various uses for the still moist grounds have been proposed (see Chapter 9); and the liquor (if not reused) will need to be treated and discharged to drain, according to local legislation. The design features of the internal filters in the columns are important, and will be numerous in their variety. A variable-speed feed-water pump (non-pulsating, multiple-throw, reciprocating) will be needed, together with a suitable heater (e.g. the normally available high-pressure steam at 200 psig or 1·380 MPa corresponds to 193°C). The extract liquor, taken off in batches, has first to be cooled to around room temperature before passing to a tank weighable *in situ*, so that the draw-off can be stopped immediately the required weight of extract has been obtained. The extract may or may not be clarified (e.g. by use of centrifuges from the wide range of suitable models from the Westphalia, Pennswalt or Alfa-Laval companies). From then on, the extract may be held in static tanks for its further processing and handling.

The use of stainless steel for all the plant in contact with coffee liquor (including the pressing of grounds) is virtually mandatory. The pressure code requirements, where applicable to columns, valves, etc., need to be stringently adhered to.

3.2. Continuous Countercurrent Screw Extractor

The alternative system to percolation batteries has again been engineered by Niro Atomizer A/S[11,30] and also by IWK of Karlsrühe in Germany. The Niro extractor described by Kjaergaard (in an industrial size, 225 kg per hour of roasted and ground coffee) consists of a narrow inclined trough mounted in a cylindrical vessel. The coffee is fed continuously into the lower end and moved upwards by two helicoidal screws (rotating at slow speed, 10–22 rph) to the upper end. The hot water feed is also at the upper end for countercurrent flow, so that the extract liquor flows towards the lower end from which it is continuously discharged. The vessel is pressurised with inert gas, and three separate and independently adjustable temperature regimes can be applied by jackets to the trough. The first at the lower end will be at 100°C; the other two will have higher temperatures, up to 180°C, the last corresponding to the feed-water temperature. The spent grounds are discharged in batches to the atmosphere. Among the claimed advantages are that less labour is involved (i.e. continuous operation, with few valves), and that there is ease of automation and little impedance to liquor flow; however, the capital costs are stated to be some 40% greater than for a percolation battery of the same capacity.

Such an extraction system can readily be used in two separate stages[30] by a first extractor operating at low pressure and a temperature of 100°C, taking a so-called atmospheric extract; and by a second extractor operating at a higher pressure and temperature fed with the spent grounds from the first extractor, to give a further extract of the remaining available soluble solids. The soluble solids concentration of the first extract can be of the order of 20% w/w (with a yield from roasted coffee of 20%), carrying the major proportion of the coffee volatile compounds; the concentration of the second extract is of the order of 12–15% w/w (with an additional yield of 20–40%), Fig. 10. The extract from the second stage can also be used, however, as feed for the first stage. Other recommendations are made for this system, e.g. the second-stage extract may be evaporated (without, necessarily, volatile compound recovery), and can be added to the first-stage extract (with or without freeze concentration) for the final drying. Such a process is described in US Patent 3,821,434 (Niro A/S, 1974).

3.3. Process Control Measurements

In respect of soluble solids, the main control is that of concentration of the extract, determined either by specific gravity or refractive index measurements. It is necessary, however, to have a correlation with actual concentration of coffee soluble solids, which may differ slightly according to the blend/roast degree/yield of the coffee being extracted (see Volume 1, Chapter 2). In non-continuous extractors, the concentration will change during the draw-off of extract, so that it is necessary to take a representative sample from the well-mixed total batch. Measurement of specific gravity with a hydrometer, or of refractive index with a refractometer on samples, or by modern in-line measuring systems, can be subject to error if entrained gases in the extract are not removed. It is necessary also to

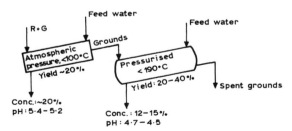

Fig. 10. Continuous countercurrent extraction using two extraction units. Courtesy of Niro Atomizer A/S, Copenhagen.

weigh accurately the batch drawn in a percolation battery, or the flow rate in continuous extraction, together with that of the incoming roast and ground coffee and its moisture content (cycle basis for a percolation battery) in order to obtain the yield figure. Samples for volatile compound content are examined by GC methods as required, for which numerous techniques are now available. The acidity of extracts is also often monitored (pH and titratable acidity, see Volume 1, Chapter 8); and the sediment content, traditionally carried out by the use of paper filter discs and their comparison with standards, is important. In order to maintain a consistent operation, it is clear that measuring devices for temperature and pressure at all required points must be accurate and well maintained.

REFERENCES

1. Perry, J. H. and various co-editors, *Chemical Engineers' Handbook*, McGraw-Hill, New York, 4th Edn, 1976, 5th Edn, 1985.
2. Schwartzberg, H. G., *Chem. Engng Progr.*, 1980, **21**, 67–85.
3. Spanninks, J. A. M., PhD thesis, Agricultural University of Wageningen, 1979.
4. Leniger, H. A. and Beverloo, W. A., *Food Process Engineering*, Reidel, Dordrecht, 1975.
5. Clarke, R. J., *Chem. Ind.*, 1976, 17 April, 362–5.
6. Loncin, M., *Proc. 8th Coll. ASIC*, 1977, 197–202.
7. Vitzthum, O. G., in *Kaffee und Coffein*, Ed. O. Eichler, Springer, Berlin, 1975.
8. Maier, H. G., *Kaffee*, Paul Parey, Hamburg, 1981.
9. Kroplien, U., *J. Fd Sci.*, 1974, **22**, 110–16.
10. Thaler, H., *Fd Chem.*, 1979, **4**, 13–22.
11. Kjaergaard, O. G. and Andresen, E., *Proc. 6th Coll. ASIC*, 1973, 234–9.
12. Thaler, H. and Arneth, W., *Z. Lebensm. Unters. Forsch.*, 1969, **140**, 101–9.
13. Sivetz, M. and Desrosier, N. W., *Coffee Technology*, AVI, Westport, Conn., 1979.
14. Macrae, R. and Trugo, L. C., *Proc. 10th Coll. ASIC*, 1982, 187–92.
15. Trugo, L. C., PhD thesis, University of Reading, 1984.
16. Segall, S. and Procter, B. E., *Food Technol.*, 1959, **13**, 266–9.
17. Pangbourn, R. M., *Lebensm. Wiss. Technol.*, 1982, **15**, 161–8.
18. Pictet, G. A. and Vuatez, L., *Proc. 8th Coll. ASIC*, 1979, 261–70.
19. Bruin, S. and Spanninks, J. A. M., *Chem. Engng Sci.*, 1979, **34**, 199–215.
20. Spanninks, J. A. M., in *Progress in Food Engineering*, Eds C. Canterelli and C. Peri, Forster-Verlag, Kusnacht, 1983, 109–26.
21. Besson, A., in *Progress in Food Engineering*, Eds C. Canterelli and C. Peri, Forster-Verlag, Kusnacht, 1983, 147–56.
22. Loncin, M., in *Food Process Engineering*, Vol. 1, Eds P. Linko *et al.*, Applied Science Publishers, London, 1980, 354–63.

23. Charm, J., *Fundamentals of Food Engineering*, 3rd Edn, AVI, Westport, Conn., 1978.
24. Desai, M. and Schwartzberg, H. G., in *Food Process Engineering*, Vol. 1, Eds P. Linko *et al.*, Applied Science Publishers, London, 1980, 86–91.
25. Pigford, R. E., *Mass Transfer Operations*, 3rd Edn, McGraw-Hill, New York, 1981.
26. Bomben, J. L., Bruin, S., Thijssen, H. A. C. and Merson, L., *Advances in Food Research*, Ed. C. O. Chichester, Vol. 20, Academic Press, New York, 1973, 2–111.
27. Thijssen, H. A. C., *J. Fd Technol.*, 1970, **5**, 211.
28. Clarke, R. J. and Wragg, A., in *Progress in Food Engineering*, Eds C. Canterelli and C. Peri, Forster-Verlag, Kusnacht, 1983, 519–21.
29. Sivetz, M. and Foote, H. E., *Coffee Technology*, Vol. 1, AVI, Westport, Conn., 1963.
30. Stoltze, A. and Masters, K., *Fd Chem.*, 1979, **4**(1), 31–40.

Chapter 6

Drying

R. J. CLARKE*

Formerly of General Foods Ltd, Banbury, Oxon, UK

1. INTRODUCTION

The drying of coffee extracts to a low moisture content (2–5%) is now generally only commercially accomplished in one of two ways: spray-drying by evaporative removal of water in a current of hot air; and freeze-drying, in which the water in the extract is first frozen and the ice sublimed off under high vacuum with controlled application of heat. Some desorptive drying still at relatively low temperature will take place, however, in the later stages of freeze-drying. Drum-drying (contact with a heated cylindrical surface), as once also used for the drying of milk, is hardly ever now used. There may also be some applications for vacuum-drying (band- and shelf-), and fluidised-bed drying is often recommended for finish drying in certain instances.

Apart from flavour quality, it is important that the dried coffee products are in an attractive and convenient physical form, and are shelf stable. The products should also be free-flowing and of a bulk density that allows the dispensing of an appropriate weight of instant coffee by a small heaped spoon for the preparation of a single cup of the beverage (say 2–3 g per 170 ml of liquid, implying a bulk density of about 18–25 g cm^{-3}). As instant coffee is soluble coffee, there are normally no problems in solution in hot water, though its rate of solution should be rapid—a further factor involving the physical structure and size of the dried

---

* Present address: Ashby Cottage, Donnington, Chichester, Sussex, UK.

particles. Spray-dried powder should not be dusty, and should have particle sizes at about 200–300 μm diameter on average, unlike many spray-dried skim-milk powders. The latter powders are now generally 'instantised' or agglomerated to increase the rate of make-up with water. Spray-dried coffees have also been agglomerated since about 1968, not so much for reasons of rapid solution, but rather more for the favourable physical appearance they present (i.e. 'granules'). In the same way, freeze-dried coffees are not prepared by the older methods for drying a pharmaceutical solution, but in a particular way, such that 'granules' are directly formed, rather than fine or very irregular and non-uniform particles. The freeze-drying of coffee on a large commercial scale commenced about 1965, and has proved to be a very successful example of this drying technique compared with the gradual commercial dropping of most other freeze-dried foodstuffs after early enthusiastic ventures. It is said that there is nothing new under the sun, and indeed a Professor Leslie of Edinburgh University was carrying out a scientific study with blocks of ice in 1811; similarly, for a long time Swiss housewives must have been freeze-drying their laundry in the cold, low-humidity, mountain air.

The coffee extraction techniques described in the last chapter lead to extracts of around 25% w/w soluble solids concentration. Although spray-drying from these concentrations will provide spray-dried coffee of the required physical form, there will be a substantial loss of volatile compounds. Freeze-drying may also be conducted at these concentrations, with the added advantage of considerably higher percentage volatile compound retention. With both methods of drying, it is, however, generally recognised that the cost of removing water (especially for freeze-drying) by drying is high; therefore there has been a considerable development in the use of pre-concentration methods for coffee extracts prior to drying at lower unit water removal costs. As already foreshadowed in Chapter 5, bulk evaporation of water (as in spray-drying itself) leads to loss of volatile compounds. Thus in that chapter the various techniques for the separate removal and subsequent handling of volatile compounds from roast coffee were covered. There are other methods of pre-concentration without the problem of loss of volatile compounds. Pre-concentration methods are therefore examined in this chapter. Evaporation, of course, may be used solely for economic reasons. All these developments are reflected in numerous patents from the major instant coffee companies, and from plant manufacturers; at the same time, there has been an underpinning by fundamental studies, notably by Professor

Thijssen and his colleagues at the Technical University of Eindhoven in The Netherlands over the period 1965–1977, but also by Karel and colleagues at Massachusetts Institute of Technology, USA, Judson King and colleagues at California Institute of Technology, USA, and many others, to whom reference will be made in the following sections. In chemical engineering terms we are concerned with selective transport[1] of water over the other constituents of coffee extract, which can be achieved either by use of rare favourable phase equilibrium relationships, as in freeze concentration; or more usually by use of favourable rate effects, as in controlled drying and reverse osmosis. Where these methods fail, as is often the case in evaporation, we are forced back into recovery methods and adding back lost desired components, e.g. volatile compounds.

2. PROCESS FACTORS IN SPRAY-DRYING

2.1. Methods

Spray-drying has been and is practised in numerous industries other than the instant coffee industry, e.g. in detergent manufacture, so there is a considerable corpus of knowledge in various general texts.[2,3] Such knowledge is rarely specifically directed towards the particular needs of, and problems in, drying coffee extracts. The subject is often dealt with primarily in terms of design requirements for efficiently removing water. Seltzer and Settelmeyer[4] around 1949 and also Clarke[5] in 1963 considered the special aspects of spray-drying of food liquids, though the issues of volatile compound retention at that time were not particularly considered and only marginally understood.

In general terms, spray-drying involves four steps: the dispersion of the liquid as a spray of droplets or globules (by means of a suitable spray device with the unfortunate name of 'atomiser'); the contacting of this spray with a stream of hot air; evaporation of the water content from the spray (with simultaneous cooling of the air by abstraction of required latent heat); and finally separation of air from the dried product. There will clearly be numerous variants of design features (even for coffee extract driers), the fundamentals of which will be discussed in the following sections. It should be noted, however, that essentially cocurrent contact air-spray is used for drying coffee extracts, and that centrifugal pressure nozzles are the most favoured type of spraying device.

The first three steps in spray-drying are carried out within vertical cylindrical chambers; the fourth may be entirely external. Spray driers in any event tend to be large constructions, especially where relatively large drop sizes are to be dried as with coffee extracts. This feature has tended to lessen the amount of experimental investigative work available compared with, say, freeze-drying.

Though horizontal spray driers have been constructed, the choice in the food industry lies between a tall, relatively narrow diameter chamber (or tower) and a broader, shorter chamber, with preference for the former in the coffee industry.

Spray-drying is, by its very nature, a continuous operation and may therefore have to be married into a discontinuous extraction system.

2.2. Compositional Changes

The composition of coffee extracts has already been discussed in the previous chapter. Only minor changes are to be expected in the nature and composition of the soluble solids, although reactions (generally undesirable) which are time–temperature dependent may occur in spray-drying and indeed in freeze-drying, and again in evaporative pre-concentration, e.g. between reducing sugars and free amino groups. The issues raised by such thermally induced changes in food products have been considered in detail by Thijssen and colleagues,[6,7] i.e. temperature–time factors required for such occurrences (Maillard/Browning reactions) compared with those same factors under commercial drying/evaporation conditions.

The time involved in spray-drying itself is only a few seconds or fractions thereof, which is a distinct advantage of this method of drying; but it is also possible for dried material (which will reach outlet air temperatures) to be removed insufficiently quickly after resting on hot internal surfaces at the outlet or separating point. Correct choice of outlet temperature is therefore important, and for a given drier/final moisture content is determined by the particle size being dried. A check on the small reducing-sugar content of extract before and after drying will serve to indicate any such likely changes. With coffee powders at low moisture content, actual fusion to hot surfaces at the outlet is unlikely (see Volume 1, Chapter 2) but is an important consideration in drying chicory or mixed coffee–chicory extracts. An extreme form of reaction in the development of some 'carbonised' particles may occur in the upper, hotter levels of a tower, in which stray air currents (or incorrectly designed nozzles) are directing spray or semi-dried particles to the hot wall surfaces that subsequently drop into the finished product.

The most important change of composition that can occur is the lower content (and change in proportions) of volatile compounds in the finished dried product. As water is being evaporated, so it might be expected that the volatile components, each component to a greater or lesser degree, will simultaneously be evaporated. This does not, however, inevitably occur; there is a marked dependence on the spray-drying conditions as discussed in section 2.5. Nevertheless, it is not clear as yet (see Volume 1, Chapter 8) as to which particular components (and how much) constitute coffee flavour/aroma, so that in general the target is to retain whatever volatile compounds are present in the coffee extract. The greatest losses are likely to be those of the smaller molecule compounds, of greatest significance to headspace aroma rather than flavour. It can also be said that poor quality extracts can be improved by standard spray-drying, by loss of undesired volatiles. High levels of such acids as acetic acid are difficult to remove.

2.3. Spray Formation

Devices for forming sprays of droplets or globules with water or aqueous (and other) solutions are widely used in many industries, including agriculture. There is considerable background knowledge on their use, design and performance, shown in general reviews, e.g. references 8 and 9, and in relation to the food industry (see reference 10). Manufacturers' brochures (notably from Spraying Systems Inc. in the USA and Watson-Delavan in the UK) are also prime sources of information.

In view of the fact that most spray-dried instant coffees have particles of relatively large size (i.e. of the order of 300 μm average diameter), it follows that the droplets in any spray will also be of that order of magnitude. The travel path for drying is comparatively long at this size, and is best accommodated within a tall narrow spray-drying tower. This in turn requires that the so-called spray cone angle from the spraying device be such that, with an inevitably high velocity of ejection of droplets, the droplets do not reach the side walls of the tower before being decelerated to gravity free fall.

In spray driers for the milk industry, so-called spinning discs are widely used. The liquid is fed at low pressure on to a surface rotating at high speed so that the droplets are spun off at high velocity in a direction close to the horizontal. Milk driers therefore have a relatively wide chamber diameter, and as large particle sizes are not called for (i.e. high powder bulk density), the total drying path is relatively short with a relatively small side-wall height. These spinning discs tend to generate fine sprays (with droplet average sizes of 80 μm diameter and much less), though coarser sprays are feasible (say 100 μm).

2.3.1. Centrifugal Pressure Nozzles

The type of spraying device most favoured for coffee extracts is the centrifugal pressure nozzle (see Fig. 1), for which high liquid pressures (i.e. of the order of 690 kPa = 100 psig and much higher) are needed, supplied by a suitable pump. Even higher pressures (13·8–69 MPa = 2000–10 000 psig) are needed to obtain small droplet sizes comparable to those obtained by means of spinning discs.

These nozzles, though small in size, have an internal cyclone-shaped chamber. The liquid enters through a pipe connection to the nozzle body,

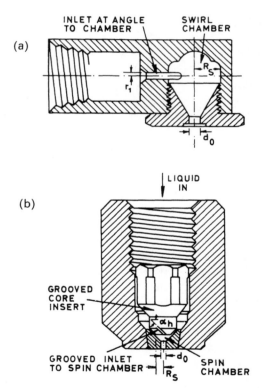

Fig. 1. Centrifugal pressure nozzles with (a) internal swirl or whirl chamber, (b) grooved core insert.

then through a small tangential orifice into the chamber where it takes up a circular path and then emerges from the bottom of the chamber through the exit orifice to form the spray. Without this chamber, the discharge of spray would be narrow and give irregular droplet sizes. The liquid actually emerges as a sheet or curtain (cone-shaped), which proceeds to break up at a short distance from the exit orifice and so provides a 'hollow' spray, of advantage in subsequently contacting the hot air. There is actually an air core at this orifice, so that the liquid passes through an annular cross-section of flow, generally of constant inner diameter for a given nozzle (independently of pressure but not viscosity) with a velocity dependent upon throughput and pressure. The entrance to a chamber is provided in a more sophisticated and expensive design by so-called grooved cores. In a given series of nozzles for a given pipe diameter, often the body is kept the same, with replaceable inserts numbered according to the orifice diameter and/or those of the grooved cores.

On exit from the orifice, the angle of spray is determined by the diameters of the internal and external orifices together with that of the inner chambers. The smaller the internal in relation to the external, the wider the spray cone angle. In effect, the governing factor is the ratio of two velocity components in the emergent liquid, one radial and the other axial.

The relationship between liquid flow rate and pressure for a given nozzle is an important performance characteristic, together with the maximum cone angle developed (usually up to 90° or 45° half-angle to the vertical), requiring some specified minimum pressure. These characteristics are readily available from the manufacturer for each nozzle, but for water spraying only. Nozzles are available in definite series of exit orifice diameter (for the simple whirl chamber). Tables of performance characteristics show a clear relationship between flow rate (Q, usually expressed volumetrically as gallons (imperial or US) or litres per minute), and the pressure drop (P, psig or metric units), such that $Q = K\sqrt{P/\rho}$ where K is a constant for the given nozzle and ρ is the density of the liquid (for water $\rho = 1$). The ratio $Q/\sqrt{\rho}$ is sometimes known as the flow number (FN) of the nozzle. It is often preferred, however, to have a constant, with the outlet orifice diameter (or radius, r_o) as a separate parameter, so that

$$Q = K_1 r_o^2 \sqrt{P/\rho}$$

In this way, in a series of nozzles with the same spray cone angle but with increasing capacity, values of K_1 are approximately constant.

Considerable studies, both fundamental and experimental, have been conducted[8,9] to determine the relationship between K_1 (and K) and the various relevant dimensions in the nozzle geometry with varying degrees of success, often unfortunately with rather small-sized nozzles. It is important also that these correlations do refer to either laminar or turbulent flow characteristics through the orifice (usually the latter in practice). In the same way there are relationships between the spray cone angle and the value of K_1 (and therefore internal dimensions).

With solutions of higher viscosity (μ_1) than water (μ_w), such as coffee extract, the volumetric flow rate will surprisingly be somewhat greater, with the small multiplying factor $(\mu_1/\mu_w)^a$, where the value of a is constant (~ 0.17). The mass flow W will also be slightly greater, due to the higher density of the liquid, i.e.

$$W = Q \cdot \rho = K_1 \cdot r_o^2 \sqrt{P} \cdot \sqrt{\rho} \cdot \mu_1^a$$

assuming the flow remains turbulent.

The higher the value of K_1, the greater will be the spray cone angle (for a given r_o^2) unaffected by low viscosities. Of further importance is the average drop size produced (which may be assessed in a number of ways, see Marshall[8]). Dombrowski and Wolfsohn[11] found that, for a range of centrifugal pressure nozzles of relevant size interest (1–724 igph and 50–350 psig), when spraying water

$$D_{32} = 332 \cdot Q^{0.3} \cdot P^{-0.5}$$

where D_{32} is the average volume–mean surface diameter (μm); and Q and P are in the above engineering units, with droplets taken fairly close to the spray nozzle. The relationship was found to be virtually independent of spray angle and of nozzle design. Little information is available for liquids of higher viscosity, though Masters[3] cites an equation of D being proportional to $Q^{0.15} \cdot \mu^{0.2} \cdot P^{-0.3}$. The effect of liquid surface tension is now generally discounted. In a series of nozzles at constant K_1 values it therefore appears that average drop size is related to the 0·3–0·6 power of r_o. Of further significance, however, is the uniformity of drop size distribution, and of weight flow distribution over the spray area, for which there is little reliable information except for water sprays. A coefficient of variation $[(D_{84} - D_{50})/D_{50}]$ of 35% at an average droplet diameter of 300 μm is a reasonable expectation of drop size distribution, where D_{84} is the diameter for 84% of the drops and D_{50} that for 50% (or average diameter). Very small drops can be produced as required, using pressure up to 10 000 psig and with nozzles of the grooved core type with a construction capable of withstanding such pressures.

As can be seen, centrifugal pressure nozzles are somewhat inflexible, in that feed rate determines pressure and so droplet size and its corresponding ejection velocity, providing problems in scale-up. Often in very large spray driers, a multiplicity of nozzles of the same size on the same pipe manifold is used, one or more of which can be withdrawn according to the capacity needed. There is some danger of nozzle blockage from solutions containing suspended matter (though coffee extract is generally pre-filtered); and of wear by abrasion of the outlet (or inlet) orifice. For the latter reason, more expensive tungsten carbide inserts can be used, though the cheaper stainless steel nozzles can be thrown away when the spray pattern is becoming affected. The major capital expense is that of the feed pump. Nozzles can handle liquids of high viscosity, say up to 450 cP, provided sufficient pump pressure is available.

2.3.2. Spinning Discs

It is more difficult than with centrifugal pressure nozzles to provide valid mathematical relationships for their performance and characteristics.[8] Spinning discs, with their variants of bowls, vanes and baskets, are necessarily more complex in construction, necessitating built-in high-speed driving mechanisms (i.e. of the order of 1500–9800 rpm), and requiring accurate hydrodynamic balancing to minimise stresses in the rotating structure. On the other hand, only a low-pressure feed pump is required.

Spinning discs have a certain flexibility in that feed rate and disc speed can be varied independently to suit a given liquid and thus provide a required range of droplet size at different outputs.

2.3.3. Other Spraying Devices

In pneumatic (or twin-fluid) sprayers, a compressible fluid such as air (or other gas) is used to disintegrate a jet of liquid as it emerges from a simple nozzle (i.e. not centrifugal). The air is usually arranged to flow at a fairly high velocity through an annular space surrounding the orifice. The spray tends to a narrow solid cone, and is particularly useful for generating small droplet sizes in a flexible manner.

2.4. Spray–Air Contact

After spraying into droplets, the liquid solution at ambient or higher temperature has to contact the stream of hot air. This can be done either cocurrently or countercurrently, with or without cross-flow, depending upon the air inlet arrangements in the chamber in relation to the position of the spraying device.[8] Strict cocurrent flow is favoured for drying coffee extract; countercurrent flow would mean that the hottest air was in contact with the driest product. This system is generally used for detergent drying. There are inevitably stray currents of air of different temperatures resulting from the evaporational drying; but the initial air can be distributed evenly over the whole cross-sectional area of the drying chamber by a suitable perforated plenum at some distance above the nozzle(s). In other designs, cocurrent air streams are concentrated from an annulus surrounding and above the nozzle (which may mean relatively high velocity), though this is only really suitable for a single nozzle installation; the air then distributes over the whole chamber area. In yet further designs, though still essentially cocurrent, the hot air is introduced tangentially to the vertical chamber so that there is an element of flow across the spray with the air descending spirally to the base of the chamber. There is here some danger of plastering of the walls of the chamber. When the droplets are widely distributed across the chamber area in free fall, the contacting of the spray with the hot air is likely to be uniform. This distribution is in consequence of the use of a 'hollow' spray, arising from a centrifugal pressure nozzle or even more so from a spinning disc; though with the latter the lower weight distribution of droplets at the centre rather than the periphery may not be satisfactory. Closer to the nozzle, the mixing of cocurrent hot air with the spray may not be very satisfactory, as a result of the effect of a local pocket of much colder air (from rapid evaporative cooling); though the use of as wide cone angles as the chamber will bear will be mitigating. Ideal cocurrent means plug flow movement of the spray with its associated air. In practice, there will be an overall profile of temperature drop from the inlet to the outlet. Some disturbance of the air pattern will also be found at the base of the tower, according to the arrangement for take-off of outlet air (see section 2.6), though this is of lesser consequence since the product will by then have been dried.

It will be noted from the following sections that in many laboratory-type drying experiments and in many theoretical calculations, a constant air temperature is assumed (or 'outlet' temperature) from perfect mixing of the spray with excess air.

2.5. Mechanisms of Water Removal

A number of different regimes for water removal or evaporation are generally recognised as occurring in a spray drier. Since the droplets contain dissolved solids, there will first of all be two distinct rates of drying, first a so-called constant-rate period, followed by a falling-rate period. The constant-rate period occurs while water evaporates freely from the surface to the surrounding hot air; the falling-rate period follows when the supply of water from the interior of the droplet can no longer sustain evaporation from the surface, which now takes place from within the droplet until all the water is substantially removed. Furthermore, the droplets (of various diameter sizes) will be ejected at high velocity (of the order of $40\,\mathrm{m\,s^{-1}}$) from the spraying device (which influences the rate of evaporation), but will be decelerated until they reach a free-falling velocity (determined by the temperature of the air outside the droplets) during the remainder of the constant-rate period (with a smaller rate of evaporation) and into the falling-rate period of drying (with even smaller rates of evaporation).

The evaporation of the water, and the consequent lowering of the temperature of the air from inlet to outlet (in cocurrent operation) occurs by combined heat transfer and mass transfer. This has been the subject of extensive studies reported in numerous texts, e.g. references 2 and 8; but importantly in recent work on foodstuff liquids by Thijssen and colleagues summarised to 1969 in reference 12, to 1973 in reference 13, and to 1979 in reference 7.

Momentum transfer is involved in following the trajectory of droplets at an angle after ejection from the spraying device to their vertical free-falling conditions under Stokes's law as discussed in Perry[2] and also by Sjenitzer.[14] Free-fall velocity is of the order of $1\cdot7\,\mathrm{m\,s^{-1}}$ for a large solid coffee particle of $400\,\mu\mathrm{m}$ in hot air at $260°\mathrm{C}$. Similar determinations can be made for other particle sizes and air temperatures, bearing in mind that dried particles may be hollow and thin-walled. Air velocities in the drier will themselves be low, of the order of $0\cdot01\,\mathrm{m\,s^{-1}}$ average.

2.5.1. Mass and Heat Balances

Marshall[8] gives a simple heat balance over a spray drier on, say, the basis of an hour of stable operation, thus: $wc_s(t_{in} - t_{out}) = q$, where w is the mass flow rate of the air; c_s is the humid heat capacity of the air (NB entering air will contain moisture); t is the air temperature; and q is the heat load based on the evaporation required in changing the moisture content from W_1 (initial, on dry basis of solid product) to W_2 (final), the sensible heat to the dry solid and to the vapour, and the heat losses. In determining sensible heats, it is necessary here to take an average temperature (t_s) for the evaporation from the spray. This temperature is assessable for constant-rate drying (when it will be lower even than the outlet temperature and that of the leaving solid), though less certainly in the falling-rate period (when it will approach or exceed that of the outlet air). The latent heat value should be taken at the evaporating temperature, which strictly speaking may have to include additional heat for bound-water desorption during falling-rate periods.

Alternatively and preferably, a heat balance may be constructed from the enthalpies of the entering and leaving solids and air, from a definite datum temperature, not therefore involving droplet/particle temperatures. Such a heat balance enables heat losses to be determined. In operating practice, a definite outlet temperature is generally set for a given tower and product by control of the burning rate of the fuel in the burner providing the hot air according to the actual load, so only the inlet temperature will vary.

Various expressions of water drying efficiency can be calculated,

$$\eta_o = \frac{t_{in} - t_{out}}{t_{in} - t_{ambient\ air}} \quad \text{overall}$$

This efficiency is higher for countercurrent than cocurrent.

$$\eta_e = \frac{t_{in} - t_{out}}{t_{in} - t_s} \quad \text{evaporative}$$

The empirical ratio of the rates of air flow to water evaporation is sometimes convenient in comparing driers, though inadequate.

A useful form of representation of the overall changing air conditions in a drier is within a conventional air humidity chart, i.e. a plot of the absolute humidity of air (in pounds or kilograms of moisture per pound or kilogram of 'bone-dry' air) against its temperature, with curves at 100% relative humidity (RH) of the air (i.e. air in equilibrium with water), and at lower values of percentage RH down to 0%, as shown in Fig. 2. Such charts also show straight adiabatic cooling lines for the air from any given air temperature (pre-calculated from mass/heat transfer data for water/air). The plot of an actual cooling line for the air in the drier can be inserted and will be determined in position by the inlet air temperature and its humidity and by that of the outlet (from observed temperature and the moisture pick-up) reflecting the heat losses. The extrapolation of this line to the line of 100% RH gives the wet-bulb temperature, or temperature of evaporation during the constant-rate period. The lowest possible outlet temperature (for the observed final

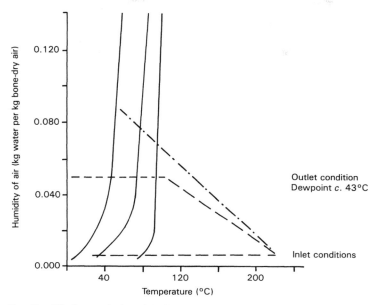

Fig. 2. Equilibrium relationship between the humidity of air (kg water per kg bone-dry air) and temperature (°C) at various relative humidities; together with typical spray-drier air heating/cooling line, from 20°C → 220°C → 100°C. —·—·—, Adiabatic cooling line with wet-bulb temperature of 54°C.

product moisture content) will be approximately obtained by extrapolation to a percentage RH figure corresponding, by its water sorption isotherm, to this moisture content. Such a plot also gives the dewpoint of the outlet air (i.e. the temperature at which its moisture content will start to condense out) by drawing a horizontal straight line from the outlet air condition to the 100% RH line. Potential condensation is of importance in the ducts and other equipment that finally convey this air to the atmosphere.

2.5.2. Rate Factors

Rate factors for the evaporation of water would be needed to determine the size of a drier for any given spray-drying operation, or the amount of water evaporated in a given size of drier. We may wish, for example, to know the required length of drying path and the drying time. In practice, the application of rate equations is difficult on account of changes in the many variables throughout the drier. Design generally follows prior experience; nevertheless some useful estimates (especially in scale-up) can be made, according to various simplifying assumptions, for both design and performance.

Rate factors are considered separately for heat transfer or mass transfer or both; though of course the mechanism of spray-drying is actually combined heat/mass transfer. Starting with the simplest case of a single pure water droplet of definite diameter, evaporating in still air and then in moving air, Marshall[8] describes the development of the relevant steady-state equations under constant conditions. Thus, for mass transfer, the rate of evaporation per unit time is given by

$$\frac{dW}{dt} = k_g A (p_s - p_a)$$

where A is the surface area of the droplet; p_s and p_a are the water vapour pressure at the surface and bulk air respectively, and k_g is the gas mass transfer coefficient across the interface for the diffusion of the water vapour through a second non-diffusing component (air), in appropriate units. The value of k_g is given, for moving air conditions, by

$$\frac{D_v \rho}{M_m x p_f} \left[2 + 0.6 \left(\frac{D_v \rho}{\mu}\right)^{1/3} \left(\frac{v_s \rho x}{\mu}\right)^{1/2} \right]$$

where D_v is the diffusivity of the water vapour in air at the surface water/air temperature, generally taken as the mean between the two; ρ and μ are the density and viscosity of the bulk air; p_f is the partial pressure of the air in the gas film; M_m is the mean molecular weight of the gas/water vapour mixture surrounding the droplet; x is the diameter of the droplet ($\pi x^2 = A$); and v_s is the relative velocity of the droplet. D_v has the dimensions [length]^2[time]$^{-1}$. It is now more general to cast this Marshall–Ranz correlation in terms of the Sherwood number in addition to the Schmidt and Reynolds numbers above, i.e.

$$Sh = \frac{k_g x}{D_v} = 2 + 0.6 (Sc)^{1/3} (Re)^{1/2}$$

where k_g will have the dimensions [length][time]$^{-1}$, and also to use water concentration units (metric) for the vapour pressure differential. Concentration units may be of various kinds, e.g. mole-fraction, weight/weight and weight/volume. The water vapour pressure or concentration in the gas film is that at equilibrium with water at the temperature of the interface.

It can be seen that in still air and when the Reynolds number is small (for small droplets), $Sh = 2$. Schmidt numbers are relatively constant with air temperature (of the order of magnitude of 0·60); whereas Reynolds numbers are not so (of the order of magnitude of 270 for a droplet of 400 μm on ejection at 30·5 m s^{-1} velocity into air at 260°C and 16 for free-fall velocity at 1·7 m s^{-1} also at 260°C, but 13 at 150°C).

For heat transfer, there is a similar but simpler equation,

$$q = h_c A(T_s - T_a)$$

where T_s and T_a are the surface and bulk temperature of the air respectively; q is the quantity of heat transferred (latent heat at surface temp, λ); and h_c is the heat transfer coefficient through the gas film equal to (again for moving air)

$$\frac{k_f}{x}\left[2 + 0·6\left(\frac{c_p\mu}{k_f}\right)^{1/3}\left(\frac{v_s\rho x}{\mu}\right)^{1/2}\right]$$

where C_p is the heat capacity of the air, incorporated in the so-called Prandtl number, and k_f is the average thermal conductivity of the gas film. Again, the correlation is now recast, with the Nusselt number

$$Nu = \frac{h_c x}{k_f} = 2 + 0·6(Pr)^{1/3}(Re)^{1/2}$$

In the SI system, k_f has the units watts m^{-1} K^{-1}, and therefore h_c has the units watts m^{-2} K^{-1}. One watt is one joule per second or 0·24 cal s^{-1}. The Prandtl number corresponds to the Schmidt number of mass transfer. By combining the heat and transfer equations, we get the surface-drop temperature (or wet-bulb temperature), i.e.

$$T_s = T_a - \frac{k_g \lambda}{h_c}\Delta p$$

in consistent units. The use of the symbol θ is now preferred to T, and certainly t, to represent temperature.

The physical properties of air are fairly well known over a range of temperatures, and a selection is given in Table 1.

The values of μ, ρ and C_p are to be used at the bulk air temperature, but k should refer to the gas film temperature and composition. The value of D_v, the diffusivity of water vapour through air, is given as 0·256 cm^2 s^{-1} at 25°C, and its value at other temperatures can be obtained by assuming $D\mu/T$ is independent of temperature (T in kelvins). According to Marshall,[8]

Table 1
Physical properties of air at selected temperatures relevant to spray drying (viscosity, density, kinematic viscosity, specific heat, and thermal conductivity) at atmospheric pressure[a]

Temperature (°C)	μ ($g\,cm^{-1}\,s^{-1}$; poise)	ρ ($g\,cm^{-3}$)	$\nu\ (=\mu/\rho)$ ($cm^2\,s^{-1}$)	C_p ($cal\,g^{-1}\,°C^{-1}$)	k ($cal\,s^{-1}$; $cm^{-1}\,°C^{-1}$)
260°	$2\cdot73 \times 10^{-4}$	$6\cdot62 \times 10^{-4}$	0·418	0·27	$1\cdot0 \times 10^{-4}$
150°	$2\cdot30 \times 10^{-4}$	$8\cdot36 \times 10^{-4}$	0·275	0·25	$8\cdot34 \times 10^{-5}$
100°	$2\cdot10 \times 10^{-4}$	$9\cdot46 \times 10^{-4}$	0·222	0·25	$7\cdot57 \times 10^{-5}$

[a] Source: Perry[2]; k values by conversion factor of $4\cdot13 \times 10^{-3}$ from Btu per hour per square foot per °F/ft units.

the gas film should be taken at a temperature halfway between that of the surface and the bulk air.

In an actual spray drier drying droplets from a spray containing dissolved solids such as coffee soluble solids, it is possible to use these equations (preferably that for heat transfer), to determine, for example, approximate lengths of drying path and travel times for a given amount of water evaporation from a given droplet size. In view of the change of value of nearly all the variables throughout the drying path, often in an uncertain manner, it is necessary to make various assumptions. In the first stage of drying, that is during deceleration, the Reynolds number is decreasing from its initial value and therefore the heat transfer coefficient; the water surface temperature can, however, be taken as constant and some average figure for the air temperature (related to the inlet temperature) can be assumed. The droplet size selected can be the average diameter in the spray. A method of calculating trajectories (time/distance) and amount of water evaporated (by increments) in this stage has been given by Sjenitzer.[14] From then on, the velocity is approximately constant in free fall, and therefore also the heat transfer coefficient. While constant rate drying is occurring, the same air temperature and surface temperature can be assumed. It is then necessary to assume some proportion of the total water to be evaporated per drop left to be evaporated in the falling rate period. In this last period, another average temperature (nearer in value to the outlet from heat/mass balance, taking into account heat losses) would have to be assumed, though again the droplet/dried particle is in free fall. Reasonable estimates can be made.

There are, however, still implicit assumptions that are or may not be correct. The droplet size is assumed to be constant. The gas film coefficient is assumed to be controlling, which, though generally true for part of the constant-rate period, will not be so for the falling rate. Though the surface drop temperature need not enter the latter stage of such a calculation using heat transfer, it will be rising from its original value. The change of control is equally evident in the mass transfer equation, where, even from early on in the drying, it is necessary to consider the decreasing internal diffusion rate of water to the surface through the increasing dissolved solids concentration, which becomes controlling. Furthermore, at the surface, the equilibrium vapour pressure of the water (or water concentration expressed in the gas phase) changes with changing soluble solids concentration in the drop and surface. Finally, only one droplet size, though average, has been considered.

Thijssen and colleagues have considered the mathematics of the real situation. As evaporation of water proceeds from the surface, so the water concentration there will fall; in fact, a water concentration gradient from the centre of the droplet will develop. The evaporation is no longer steady state (even with constant external factors), since the diffusivity of the water in the droplet is not constant and is known to vary very markedly (of the order of 10^2- to 10^6-fold) with the concentration of dissolved solids. This phenomenon is important in considering its effect on the retention of volatile compounds also present in the solution or extract being spray-dried, as will be discussed in section 2.6.1. Like the extraction of soluble solids discussed in Chapter 5, it is necessary to consider 'intra-particle diffusivity'.

Rulkens and Thijssen[15] in 1969 showed how this concentration gradient develops and the type of computerised mathematical solution necessary for the diffusion equations for water transport, even in the simple case of an infinite length of a slab of definite thickness drying in air at a constant temperature, and at a constant surface temperature. Maltodextrin as the soluble solid was used for a specific example; and a small amount of gel agent had been added to prevent convective movements. By Fick's first law of diffusion, the flux of water (relative to stationary coordinates) is given by

$$N_w = -D_{wd}\frac{\partial C_w}{\partial Z}$$

and related to the movement of soluble solids molecules

$$N_w^d = -\frac{D_{wd}}{1 - C_w/\rho_w}\frac{\partial C_w}{\partial Z}$$

where D_{wd} (cm² s⁻¹) is the binary diffusion coefficient of water–dextrin; Z is a distance at right angles to the gas–liquid phase; C_w (g cm⁻³) is the water concentration; and ρ_w (g cm⁻³) is the density of water. At the surface, N_w^d equals the water flux, F_w (g cm⁻² s⁻¹), i.e.

$$N_w^d = F_w = k'_w(C'_{w,0} - C'_w)$$

where k'_w is the mass transfer coefficient in the gas phase; and C'_w and $C'_{w,0}$ are the water concentrations in the bulk gas phase and that in the gas phase at the interface (which is in equilibrium with $C_{w,0}$, the water concentration in the slab also at the interface). It is necessary to have a relationship between $C'_{w,0}$ and $C_{w,0}$ at the slab temperature, which is given by the known water sorption isotherm for malto-dextrin solutions. Though usually the isotherm is given in terms of water activity and per cent water concentration in the solution, the relationship used is of the type

$$C'_{w,0} = \rho_w[A_w - B_w \exp(-K_w C_{w,0}/\rho_w)]$$

where ρ_w is the density of water, and A_w, B_w and K_w are functions of the slab temperature. The relationship between C'_w and C_w is that for pure water. It is also necessary to have a relationship between the diffusion coefficient D_{wd} and C_w, which is again of exponential form as described on page 165. By computation by time intervals, Rulkens and Thijssen were able to show the changing water distribution pattern from surface to interior with time, corresponding to that found in the earlier experimental work of Menting[16] with the steep concentration gradients that developed. The evaporation rate in terms of final/initial water concentration as a function of time can also be calculated for different drying times.

By use of Fick's second law of diffusion,

$$\frac{\partial C_w}{\partial t} = \frac{\partial}{\partial z}\left(D_w \frac{\partial C_w}{\partial z}\right)$$

the surface concentration $C_{i,w}$ can be calculated and plotted against drying time for different initial water concentrations and for different values of k'. Thijssen[12] provides a diagram showing that the surface concentration of a drying slab will be reduced from 0·75 g cm⁻³ (~75% w/w) to 0·15 g cm⁻³ in about 0·1 s with a mass transfer coefficient value (k'_g) of 50 cm s⁻¹; and in about 0·01 s starting with a surface (and average) concentration of 0·50 g cm⁻³. The surface concentration has a particular significance in the retention of volatile compounds.

Thijssen[7] further discusses a 'short-cut' method for calculating the drying time of homogeneously shrinking particles, which was developed by Schroeber, where he states that the initial flux of water is dependent upon the external air temperature and initial water (solids) concentration. The higher the initial flux, the steeper is the water concentration gradient, so that the interfacial water concentration falls to any given value in a shorter time, and at a higher average water concentration.

In a further application of this short-cut method, the spray-drying of droplets containing dissolved solids is considered. The air temperature and humidity can be assumed constant with well mixed air; or can vary along the drier as in actual cocurrent spray-drying. In the latter case, air conditions have to be related to the moisture content of the particle at any time by heat and mass balances. In each case however, unlike the drying slab previously considered, the particle temperature and interfacial water concentration now vary continuously and are dependent variables. The mass transfer and heat transfer equations have to be included, i.e. those of the Ranz–Marshall correlations already noted. With droplets or particles in free fall, both the Sherwood and Nusselt numbers are essentially equal to 2 for small particles (say 50 μm) since the Reynolds numbers will be low. However, for particles of the order of 400 μm, Reynolds numbers will be of the order of 16, so that the additional mass transfer figure is of the order of a further 2 units. The results of calculations are presented in this paper, showing the relationship between drying time (0–1.5 s) and moisture content (1.5 to 0.05 kg water per kg dry matter) for spray-drying a maltose–water droplet of 25 μm for two air outlet temperatures (120 and 80°C), though at constant air outlet humidity, in well mixed air. The drying rate was faster at 120°C than at 80°C. These results were stated to compare very favourably with those from a rigorous calculation. In actual spray-drying, it is also necessary to assess the drying time and moisture content during the intial period of deceleration, when the evaporation rate can be high (say up to 30% for large particles at high initial velocity but less for smaller).

The necessary data for the physical characteristics of malto-dextrin and maltose solutions (i.e. water diffusion coefficients and water sorption isotherms) are available; they are similar to those for coffee extracts but not identical, as discussed in Volume 1, Chapter 2, and approximately indicated in Table 2.

The relationship between diffusion coefficients and water concentration at a given temperature is of the general type

$$D_w = \exp[A + B(1 - x_w)]$$

where x_w is the mole fraction of water present, and A and B are positive

Table 2
Diffusion coefficients (approximate) and activities for water in various solutions at different temperatures[a]

Solution	Water concentration (% w/w)	At 25°C D_w ($cm^2 s^{-1}$)	A_w	At 60°C D_w ($cm^2 s^{-1}$)	A_w
Coffee extract	50	3.0×10^{-6}	0.95	—	0.95
	40	1.5×10^{-6}	0.90	—	—
	15	5.0×10^{-8}	0.68	—	—
Malto-dextrin	50	1.9×10^{-6}	0.975	—	0.975
	40	1.0×10^{-6}	0.93	—	0.93
	15	1.9×10^{-8}	0.75	—	0.80
Maltose[b]	50	1.6×10^{-6}	0.94	3.3×10^{-6}	0.94
	40	1.0×10^{-6}	0.91	2.3×10^{-6}	—
	15	5.9×10^{-8}	0.65	2.2×10^{-7}	—

[a] Data source: Thijssen[12] and Kerkhof and Thijssen[25] graphical plots; Menting et al.[17]
[b] Kerkhof and Schroeber;[18] calculated from equations.

or negative constants. To express the effect of temperature, another term is needed, i.e.

$$D_w = \exp\left[A' + B'x_s - \left(\frac{C + Dx_s}{RT_k}\right)\right]$$

where x_s is the mole fraction of dissolved solids (i.e. $1 - x_w$), C and D are further constants, R is the gas constant in appropriate units (1.987×10^{-3} kcal mol^{-1}), and T_k is the absolute temperature (see Volume 1, Chapter 2). Care has to be taken in respect of the type of diffusion coefficient, whether binary, self-diffusion, etc.[17,18] The data for malto-dextrin solutions were obtained by Menting, who correlated his experimental data over a limited range as

$$\frac{D_w}{D_w^0} = 5.0 \exp(-1.6\rho_w/C_w)$$

where ρ_w is the density of water ($= 1$); C_w is the concentration in g cm^{-3}, which requires information on the density in relation to a % w/w or mole-fraction figure; and D_w^0 is an extrapolated figure for pure water, equal in value to 3.4×10^{-6} cm^2 s^{-1}. This latter figure is lower than the known self-diffusion coefficient of water at 2.2×10^{-5} cm^2 s^{-1}. Diffusion coefficients at higher temperatures do not appear to be available, but are governed by $D = D_0 \exp(-E/RT)$. The grade of malto-dextrin is believed to be 20DE of an approximate molecular weight of 500.

For solutions of coffee soluble solids, data are only available in terms of % w/w or kilograms per litre, and appear to be lacking for temperatures other than 25°C. Some equivalent molecular weight figure would have to be used for the mole-fraction expression of the equation.

Comprehensive data for maltose have been obtained and correlated by Van Lijn, quoted by Kerkhof and Schroeber.[18] Maltose has a known molecular weight of 342. The constants $A(A')$, $B(B')$, C and D are known (see Volume 1, Chapter 2).

The relationship between water activity and water concentration at a given temperature is of the type

$$\log(A_w/x_w) = K_2 x_s^2 + K x_s^3$$

where A_w is the water activity, x_w and x_s are mole fractions as before, and K_2 and K are constants. Alternatively, in exponential form

$$A_w = x_w \exp(K'_2 x_s^2 + K' x_s^3)$$

At higher activities (close to 1·0), the effect of temperature is small. Again see Volume 1, Chapter 2, for values of these constants for particular aqueous solutions.

2.6. Mechanisms of Volatile Compound Retention
2.6.1. The Selective Diffusion Concept

This concept was developed and subjected to experimental testing by Thijssen and his colleagues from 1965 onwards in conjunction with the studies already discussed in section 2.5.2 for the evaporation of water alone; see reviews in 1969,[12] 1973[13] and 1979.[7]

Volatile compounds, including those in coffee extract, have a wide range of relative volatilities. They are generally very high relative to water, especially at infinite dilution, and will therefore evaporate from a water surface at a much faster rate than water. There will be a very rapid depletion of those substances at the surface, so that thereafter their loss becomes controlled by their convective and molecular transport (be it in a slab of aqueous solution, in droplets or in some other shape). In the absence of convective movement, the transport of both volatile compounds and water will be governed by molecular diffusion. The molecular diffusion coefficients of volatile compounds (D_j for a given volatile component j) markedly decrease (like water itself, but even more so) with decreasing water concentration and with decreasing temperature. It has already been noted (section 2.5.2) that, as a result of water evaporation, a water concentration gradient develops within the rigid drying body. This concentration gradient in turn sharply decreases the diffusivity of the volatile components. It is evident, therefore, that conditions within the drying body can be created for rapidly reducing the diffusivity of the volatile components, and thus their capacity for escape at the evaporating surface. The volatile compound molecules have to become less mobile than the water molecules (for their effective retention). The relative volatility of the volatile components then ceases to have any significance; the controlling parameter is the ratio of volatile compound to water diffusivity, D_j/D_w in the solution. For effective retention of volatile substances, steep and rapidly developed water concentration gradients have to be established within the drying body, leading to the formation of a 'skin' or 'crust' at the surface, which is impermeable to the volatile substance. The establishment of these water concentration gradients has been shown to be determined by the characteristics of the drying air (temperature, humidity, velocity, etc.), but also of the drying body itself. The beneficial effect on volatile compound retention of initial high soluble-solids concentrations has long been observed in practice and finds a ready explanation in this selective diffusion concept, i.e. by rapid formation of steep water gradients.

Experimental information is also available for values of the diffusivity of a number of volatile compounds (notably acetone) in malto-dextrin and maltose solutions and in coffee extracts of different water concentration and temperature, as approximately indicated in Table 3. Diffusion coefficients are inversely related to the logarithm of molecular size (diameter), and less exactly to molal volume and molecular weight. The Wilke–Chang equations[2] allow estimates of diffusivity of one compound from that known for a reference compound in water and in the same known solutions at the same temperature. The diffusivities of volatile compounds in coffee can range over a factor of 2–3.

It is important to note that the foregoing applies only to diffusive movements of both the volatile compounds and the water, which occurs in a rigid body (e.g. gelled slabs, or droplets at high viscosity); mechanisms of other volatile compound loss which may occur are discussed in section 2.6.2. 'Diffusive' loss will take place during the constant rate period of drying, and less so subsequently. Thijssen[19] discusses also semi-qualitatively a so-called critical value for the surface water concentration, when the ratio D_j/D_w has become sufficiently low (say 0·01) for volatile loss at

Table 3
Diffusion coefficients for acetone in different aqueous solutions at two different temperatures[a]

Solution	Concentration of water (% w/w)	Diffusion coefficient (D_j, cm^2 s^{-1}) at 25°C	at 60°C
Coffee extract	50	$1{\cdot}3 \times 10^{-6}$	—
	40	$7{\cdot}3 \times 10^{-7}$	—
	15	$1{\cdot}0 \times 10^{-8}$	—
Malto-dextrin	50	$1{\cdot}3 \times 10^{-6}$	—
	40	$4{\cdot}0 \times 10^{-7}$	—
	15	$2{\cdot}9 \times 10^{-10}$	—
Maltose	50	$1{\cdot}4 \times 10^{-6}$	$3{\cdot}2 \times 10^{-6}$
	40	$6{\cdot}8 \times 10^{-7}$	$1{\cdot}7 \times 10^{-6}$
	15	$5{\cdot}0 \times 10^{-9}$	$2{\cdot}5 \times 10^{-8}$

[a] Data source: references 12, 17, 18 and 25, as Table 2.

the surface to have virtually ceased. The actual critical value (like D_j/D_w) will depend upon the particular volatile component, the temperature of the evaporating droplet, and the nature of the soluble solids (affecting diffusivities). A typical critical moisture content figure of 15% w/w is quoted, though this depends upon the particular volatile compound and the temperature as indicated from the approximate figures in Table 4 for malto-dextrin solutions.

In considering the values of D_j/D_w for different water concentrations for acetone in different solutions and extracts, it appears that lower surface water concentrations have to be reached for D_j/D_w values as low as 0·01 in coffee extracts as indicated approximately from Table 5,

Table 4
Critical moisture contents for three different volatile compounds[12,25] in malto-dextrin solutions ($D_j/D_w = 0{\cdot}01$)

Volatile compound	Temperature (°C)	Critical moisture content (% w/w)
Acetone	25	7
	0	10
Methanol	25	7
	0	15
n-Pentanol	25	18
	0	30

Table 5
D_j/D_w values for acetone in various solutions at different water concentrations[a]

Water concentration (%)	Malto-dextrin (aq.) 25°C	Coffee extract 25°C	Maltose (aq.)	
			25°C	60°C
50	0·53	0·6	0·90	0·97
40	0·40	0·5	0·68	0·73
15	0·02	0·2	0·08	0·11

[a] Calculated from Tables 2 and 3.

deduced from Tables 2 and 3. The data may not be so reliable for coffee extracts compared with those of the other solutions.

Except for maltose solutions, data for higher evaporating temperatures do not appear to be available, though these will be experienced in spray-drying. The variation of D_j/D_w with temperature ($T°$ absolute) will be given by an equation of the form

$$\log D_j/D_w = \text{Constant} - \text{Constant}/T$$

so that D_j/D_w decreases with decreasing temperature.

In the same way, a critical time (t_c) can be defined for the surface water concentration to become sufficiently low to cause negligible loss of volatile compound(s). The dimensionless Fourier number (Fo) incorporates this drying time, the diffusion coefficient D_j (averaged or taken as a constant down to the critical moisture content) and the relevant dimensions of the drying body (e.g. radius of the drying droplet), i.e. $Fo = D_j \cdot t_c/r^2$. This number is known to be related to the percentage volatile compound retention (see section 2.6.3), so that the lower the Fourier number for whatever change of variable (D_j, r, or t_c, dependent upon the external drying conditions), the higher the percentage retention. On this basis, the process variables are chosen to keep Fo minimal. This relationship is of at least qualitative value.

The effect of a further dimensionless number, the Biot number, in relation to volatile compound retention is discussed by Menting et al.[21] Loss of volatile compounds will only be slight if, at the onset of drying, the value of the dimensionless group $Bi_{wo} K_w$ is sufficiently large (~ 22), where Bi_{wo} is the Biot number for water transport, and K_w is the ratio of the equilibrium water concentration in the gas phase and the water concentration in the solution to be dried. The Biot number reflects the relative significance of the mass transfer coefficient in the gas phase and that of the water diffusion coefficient, i.e. $Bi_{wo} = k'_w/D_w$.

2.6.2. 'Non-diffusional' Loss

There are a number of 'provisos' in the application of this concept. In spray-drying liquids, droplets are initially formed by shearing and are ejected at high velocity; in conjunction with the effect of high drag forces on deceleration, internal movements are set up within the liquid droplets, i.e. convective movements overpowering simple molecular diffusion, so that, in this brief period, evaporative rates of loss of both water and volatile components will be high but none the less calculable. Furthermore, unless the mixing of air and spray is ideal, a high local concentration of droplets, especially of small diameter from a nozzle with a narrow cone angle, is likely to cause initial air cooling and high humidity, which then serve to depress evaporative rates and thus the desirable steep water gradients. A further local effect occurs close to the spray device, say up to 15 cm vertical distance, where the expanding film from a nozzle orifice is formed and then breaks up into droplets.[1,22] King found quite high losses in model experiments with 40% w/w and 1% sucrose solutions with added volatile compounds. For example, with a 40% solution containing ethyl acetate (100 ppm), correlations from the model work predicted the loss of 10% in the expanding film, 7–15% during drop formation, of the volatile compound from a fan-shaped nozzle (operating at 690 kPa pressure = 100 psig) into air at 150–160°C and a liquid feed temperature of 43°C (corresponding to the wet-bulb temperature). The liquid flow rate was 6·1 litre h^{-1} (water equivalent), and the air flow rate was 100 kg h^{-1}. The further loss during the internal circulation period was 6%, and that of the diffusion period (up to 16 cm from the nozzle) was 31%. In this total period, the percentage of water evaporated was still only about 5%. King found gas phase control of the aroma loss in the expanding film, but more liquid phase control further from the nozzle.[1,22]

Another potential source of additional loss, recognised by Thijssen, is in the so-called 'ballooning' or puffing of dried droplets. The inflation (and also shrivelling) of particles has also been examined by King,[1] who considers that desorption of dissolved air (as a result of increase in drop temperature) within the droplet is required for internal bubbles to form, which then have a mole-fraction content of water approaching unity, and expand when and after the temperature rises above 100°C. Expansion occurs earlier and to a greater extent for drops that are heated to a higher temperature during their fall. Even if the feed liquid contains little or no dissolved air, absorption of air can take place close to the spraying device, though this should be minimised with high feed temperatures. This ballooning phenomenon is especially noticeable when spray-drying extracts of relatively low concentration (and when foamed); the resultant powders also have a low bulk density (say below $0.20\,\text{g cm}^{-3}$), associated with the use of high inlet temperatures. When the expansion is sufficiently powerful, craters can form on the surface and crack so that volatile compounds can be lost from both sides of the particle wall. An estimate of the loss is typically 10–20%. It can therefore be seen that the additional losses that can occur, before the strictly diffusional loss according to the selective diffusion concept is applied, are of the order of 10–30%.

Yet a further consideration is the presence of an oil phase emulsified in the aqueous phase, which influences volatile compound retention. Such an oil phase can be present in many coffee extracts (say to 0·1% w/w on a solids basis), though centrifugal separations may reduce this level. Coffee oil may be added back to favour volatile retention, as proposed in several patents. Zakerian and King[22] have also considered this subject, with the same model conditions used for studying volatile loss close to the nozzle. They concluded that an added oil phase can be very beneficial for volatile compound retention, but will also increase both average drop size and its spread.

Kerkhof[23] distinguished between the case of volatile components homogeneously dispersed in the solution being dried and the less frequent case of dispersion when the components have exceeded their solubility in the solution and where surface active agents can determine the degree of dispersion. Experimental work was not complete, but some differences in percentage retention were predicted.

2.6.3. Rate Factors

The full mathematical treatment of this selective diffusion concept involves considerable complexity, as may be seen from an examination of the series of scientific papers and reviews describing this work in its developing stages from 1965. Some changes in symbols, interpretation and presentation should be noted in this period.

Consideration starts with the simple case of the diffusive drying of slabs of aqueous solutions (made rigid with small quantities of gum to minimise convective movements of water and volatile substances) under isothermal conditions (i.e. drying air temperature essentially that of the outgoing air, and constant surface temperature). It proceeds then to the drying of spherical droplets (both non-rigid and rigid through inherent viscosity), first with completely mixed air (again a single air temperature), and secondly with the more real situation of limited cocurrent (or countercurrent) air usual in commercial dryers. The temperature of a drying droplet can no longer be regarded in either case as constant, but can be uniform through the droplet.[18]

In addition to the water diffusion–evaporation equations given in section 2.5.2, Rulkens and Thijssen[15] give similar equations for the transport of the volatile compounds in the drying of an infinite slab, i.e.

$$N_j^d = -D_j \frac{\partial C_j}{\partial Z} + \frac{C_j D_j}{\rho D_{jw}} N_w^d$$

where j refers to the volatile compound (actually a, for acetone, in a worked example), and d refers to the malto-dextrin. There are similar equations (to that of water) for the evaporative flux of the volatile compound, for the relationship between the concentration of the volatile component in the gas (air) phase, and the concentration of water in the solution phase. Care has to be taken in the selection and significance of the various diffusion coefficients, either binary, multi-component or mutual diffusion, as further discussed by Kerkhof and Schroeber.[18] In the same way as for water, the distribution of the volatile compound in steep gradients in the slab with time was obtained by a computerised solution of the basic equation. After a time, the flux of volatile compound at the surface is nearly zero, and the slab is losing only water.

Further mathematical treatment involves the use of Fick's second law of diffusion, i.e.

$$\frac{\partial C_j}{\partial t} = \frac{\partial}{\partial z}\left(D_j \frac{\partial C_j}{\partial z}\right)$$

which describes the concentration of the volatile component (j) at a distance (z) from the interface in relation to time t. There is a well known solution[21] to this equation for the ratio $\bar{C}_{jt} - C_{ji}/C_{j0} - C_{ji}$ where $\bar{C}_{jt}$ is the average concentration after time t, and the subscripts 0 and i refer here to the initial and interfacial (or surface) concentrations in terms of the Fourier number, $Fo = D_j \cdot t/\text{distance dimension}$. The distance dimension depends upon the shape of the drying surface, and for a droplet is r^2.[21,24] By the selective diffusion concept and resulting from the high relative volatilites of the volatile component ($\alpha > 10$) and the rapid rate of evaporation of water (high k_g' value), C_{ji} is zero almost from the start of drying, so that the residual ratio C_{jt}/C_{j0} is the actual fractional retention of volatile components in time t. The relationship between $\bar{C}_{jt}/C_{j0}$ and Fourier number for a spherical particle is of the type (approximation to one exponential factor)

$$\text{Volatile retention} = 1 - 6(Fo/\pi)^{1/2} \quad \text{for } Fo < 0.022$$

or

$$\frac{6}{\pi^2}\exp(-\pi^2 Fo) \quad \text{for } Fo > 0.022$$

and is conveniently used when graphically presented, e.g. a plot of Fo against log(volatile retention). A fractional retention of 0·90 or better means Fo has to be less than 0·01, and 0·37 retention corresponds to 0·05. In the Fourier number t is taken as t_c, the critical time at which changes in the water concentration have just formed an impermeable layer to further volatile compound loss. The critical time is dependent upon the effect of external air conditions on the rate of water evaporation, governed as we have seen by $k'_g(C'_{wi} - C'_w)$. Assuming for the moment a constant value of D_j, Thijssen[12] illustrated the effect of various changes in the other constants, e.g. droplet size, and showed that retentions up to 90% should be readily obtainable during spray-drying in those stages where a molecular diffusive mechanism was applicable, especially with an initial water concentration of 50% w/w. With a Sherwood number of 2, retention was calculated to be independent of droplet size, though not when spray-drying from a malto-dextrin solution at 75% w/w. The calculated retentions were also of this order (80–90%) even when spray-drying from the less concentrated solution, which is not experienced in practice (see page 178, suggesting that the 'non-diffusive' losses (section 2.6.2) were playing a very predominant role. However, lesser retention in the diffusive drying is reported in more rigorous mathematical treatment, see page 175. In practice, D_j cannot be regarded as a simple constant, so that either a rigorous mathematical solution is required or a short-cut method must be used, as described below.

The significance of the ratio D_j/D_w can be seen from the simple mathematics of the equation covering diffusion in the liquid phase and evaporation in the gaseous phase.

In the paper by Kerkhof and Thijssen,[25] also reported by Thijssen[7] and by Kerkhof,[23] the diffusional loss of volatile compounds in actual spray-drying was considered quantitatively predictable by a short-cut method from simple slab drying experiments, to which an estimate of the 'non-diffusional loss' was to be included to give the total loss. The basis of the method is to obtain a realistic Fourier number for the drying conditions, which includes an effective diffusion coefficient ($D_{j,\text{eff}}$). This Fourier number can then be used in the relationship between volatile compound retention and Fourier number, already discussed. The simplification involved is to define the constant rate period of drying (during which the major diffusional loss is occurring) as the length of time during which the interfacial water activity is higher than 0·90. The corresponding water concentrations can be seen from the water sorption isotherms for the particular solution or extract; e.g. above 40% w/w for a coffee extract (and maltose), 30% w/w for malto-dextrin, compared with 48% w/w for sucrose, at 25°C (though these figures are little affected by temperature). These values appear higher than the previously described critical moisture contents. Fluxes of water under various slab drying conditions have to be obtained by experiment to give the value of various constants and so an effective diffusion coefficient and Fourier number. $D_{j,\text{eff}}$ was found to be solely dependent upon the initial water content and the surface temperature during the period of change of activity down to 0·9, and its value was equal for slabs and droplets. The length of the 'constant' activity period is dependent upon the initial water content, air temperature and humidity and droplet size.

In the paper by Kerkhof and Schroeber[18] the multi-component diffusion coefficients and all other variables are considered in much more rigorous detail, and again computerised solutions of the differential equations are presented. With necessary changes of coordinates, use of dimensionless expressions for all variables and correct choice of boundary conditions, the actual percentage retention was calculated for the spray-drying of droplets of different diameters and of given initial water content, in air of a given temperature and humidity (i.e. generally well-mixed air, assumed for a constant drying air temperature); including, however, the effect of changing droplet temperature and of size. A model solution of acetone in maltose was taken, with its known physical properties. Retentions were calculated separately for the first stage in which the droplets are ejected at high speed from the nozzle (100 m s^{-1} into air at 140°C), and when internal circulation movements are set up in the droplet; and for the second stage, with rigid droplets falling in air at different temperatures (where $Sh = 2 = Nu$). The equations used both for the transport of water and volatile compound molecules, which must be solved simultaneously, are given in reference 18 (also in reference 26).

Final retentions were shown to increase markedly (a) with increase in initial solids concentration, e.g. only 4% at 20% w/w, to 68% retention

Table 6
Factors maximising percentage volatile compound retention[13,18]

Item No.	Factor
1.	High initial dissolved solids concentration (or low water concentration); say above 40% w/w for coffee solubles.
2.	High molecular weight of the volatile compound (i.e. low diffusion coefficients; but generally also have high relative volatility). NB Coffee extract contains volatile compounds with a wide range of volatilities and diffusivities.
3.	Absence of convective circulation in drying droplets, assisted by items 1, 8 and 9a, but not 4a.
4.	High gas phase mass transfer coefficients (k'_g for the water molecules), dependent upon (a) droplet velocity (which varies from ejection to free fall); (b) droplet initial diameter; (c) physical properties of surrounding air, and diffusion coefficient of water vapour in the gas film, reflected in Sherwood, Schmidt and Reynolds numbers. Overall, higher air temperatures increase the value of k'_g (Schmidt number relatively constant), but can influence 'ballooning'. NB Initial droplet velocity and size are interrelated with centrifugal pressure nozzles. Overall effect of droplet diameter not certain, but increasing diameter probably favourable, especially with the more dilute solutions.
5.	High driving force differential for water evaporation (that is, between drying water surface and bulk drying air), and therefore low external air humidity required.
6.	Low liquid diffusion coefficient of water in drying droplet; D_w decreases with increasing water concentration and increases with temperature, and decreases with increasing molecular weight of soluble solids.
7.	Overall effect of droplet evaporating temperature probably small, as consequence of two opposing effects.
8.	High molecular weight of dissolved solids. NB Coffee extracts relatively constant in average molecular weight, though will be affected by blend, roast colour and extraction yields.
9a.	Increasing feed temperature, especially at higher dissolved solids concentrations.
9b.	Decreasing feed temperature at low dissolved solids concentrations (increased viscosity effect).
10.	Increasing outlet temperature in mixed air flow only (up to a maximum, when 'ballooning' can occur).

at 55% w/w solids concentration; (b) with increase in air outlet temperature from 75 to 140°C; less so (c) with feed temperature from 20 to 80°C; and (d) with initial droplet size (10 to 100 μm) for a high initial velocity; and to decrease with (e) increasing outlet air humidity. In each case, the other relevant variables were held constant. It was found that volatile compound losses ceased when the surface water concentration was reduced to a value of about 25% w/w.

Results of computation using modified equations were also presented by Van Lijn et al.[27] for droplet heat/mass transfer under spray-drying conditions, which, however, included the case of perfectly cocurrent flow with different inlet air and outlet temperatures (i.e. 175, 200 and 250°C; 70, 90 and 120°C). However, the droplet considered was only 20 μm in diameter, with an initial solids concentration of 50% w/w (maltose), with a Nusselt number of 2. Of interest are the findings that in cocurrent air flow the droplet temperature rises quickly to a figure above the outlet temperature (especially at the higher inlet and lowest air outlet temperature), but falls again before the end of drying; the critical Fourier number to reach a surface water concentration of 15% decreases with inlet temperature, and is little affected by outlet temperature; and that compared with mixed or flow drying, the drying times are shorter though the droplet temperatures are generally higher.

Summarising the qualitative effect of basic variables in maximising percentage volatile compound retention, Table 6 indicates the main trends, covering both the deceleration and free-fall period in spray-drying (mixed air and cocurrent).

2.6.4. Experimental Trials

Various experimental trials on the drying of slabs[15,28] with observed retentions of volatile components from GC analysis have indicated the essential correctness of the foregoing thesis and equations, though generally with a reduced number of variables (e.g. constant temperatures). Experimental verification under spray-drying conditions, with further interconnected variables, is rather more difficult; but helps the choosing of external variables to optimise retention. The external measurable variables in a given spray drier of fixed dimensions are indicated in Table 7.

Rulkens and Thijssen[29] described their experimental work with a small spray drier in assessing the effect of a number of variables, as given in Table 8. Two types of malto-dextrin DE20 (but of different viscosity at the same solids concentration in solution) were used; the volatile compounds carried at 1000 ppm were acetone and various alcohols in a homologous aliphatic series.

The measured volatile retention includes losses from all sources (i.e. including ballooning that takes place close to the nozzle and any internal circulation effects). The most significant effect was that of soluble solids concentration, in which, after about 43% w/w, the retention of acetone

Table 7
External variables in spray drying[a]

Item	Variable
1. Spraying device	(a) operating pressure (with centrifugal pressure nozzle)
	(b) nozzle geometry determining:
	(c) droplet size/initial ejection velocity/cone angle/spray concentration
2. Liquid feed	(d) temperature at nozzle
	(e) dissolved solids concentration
	(f) flow rate
3. Air flow	(g) inlet temperature and humidity
	(h) outlet temperature
	(i) flow rate and velocity within drier
4. Product	(j) final moisture content
	(k) volatile compound retention
	(l) bulk density/colour/screen analysis

[a] For fixed dimensions of drier (diameter and vertical drying path). Only items (b), (d), (e), (f), (h) and (i) are truly independent variables.

jumped markedly from only 25% to nearly 80% when an inlet temperature of 210°C and a feed temperature of 85°C were used, together with the high-viscosity malto-dextrin solution at a feed rate of $0.045\,m^3\,h^{-1}$. The final moisture content was not given, nor average particle size, which would vary with the solids concentration, nor the bulk density, which could be quite high (i.e. c. $0.3\,g\,cm^{-3}$ packed). Higher inlet temperatures decreased the percentage retention figure; outlet temperatures are not given. With low feed temperature (20°C), and therefore even higher viscosity producing somewhat larger droplets, the percentage retentions are much lower (30%) at all inlet temperatures at high solids concentrations, but somewhat higher at lower concentrations (35%).

A further significant effect of maintaining all spray-drying conditions constant but varying the nature of the volatile component was the marked increase in retention of higher homologue alcohols (methanol to n-pentanol) with their inherently lower diffusion coefficients yet much higher relative volatility (α_j^∞, i.e. at infinite dilution, see section 2.6.1) in conformity with the selective diffusion concept; and indicating that this is the prime mechanism occurring, though an unknown and varying percentage of any volatile loss will be due to volatility factors.

Table 8
Pilot spray drier variables in experimental trial[29]

Item No. from Table 7	Range of values[a]
1. Jato centrifugal pressure nozzle	(a) and (b) characterised for a 50% w/w malto-dextrin solution, by:
	Feed rate ($m^3 h^{-1}$) — Operating pressure ($kg\,cm^{-2}$)
	0·035 — 20
	0·050 — 40
	(c) not given
2. Liquid feed	(d) 85 and 20°C
	(e) 35–55% w/w
	(f) 0·02–0·100 $m^3 h^{-1}$
3. Air flow	(g) 290–250–235–210°C
	(h) not given
	(i) 600–900 standard $m^3 h^{-1}$
4. Product	(j) measured but not given, presumably all below 5%
	(k) measured, range 5–80%
	(l) Bulk density (packed) measured, 0·1–0·5 $g\,cm^{-3}$, varying essentially and inversely with inlet temperature

[a] Spray drier dimensions given as diameter, 2·5 m.

2.7. Fines Separation

As a consequence of the range of droplet sizes produced from the spraying device, there will be a corresponding range of particle sizes in the dried powder at the bottom of the tower, though the relationship between the two in respect of average size and range (expressed as a coefficient of variation) will depend upon the drying conditions (degree of particle contraction/expansion). In either case, the cumulative weight percentage for a given particle size will give a reasonably straight line when plotted as a probability against size (aperture of screen used for analysis).

This total powder has to be separated out in cocurrent drying from its conveying outlet air; virtually the whole size range could constitute the final product (for sale), but in practice a fraction is taken out (so-called fines). Usually a cut-off point is decided (i.e. a given small percentage only at a given particle size in μm is to be left in the product, say 5% at 200 μm and less. Cyclones external to the drier can be used providing a single discharge point; but a primary separation can be made within the drier by a construction of its base whereby the outlet air is taken off in a duct at right angles to the vertical tower. This primary separation can be adjusted so that a coarse product corresponding to the desired finished product falls out by gravity from an orifice of the cone-shaped base of the tower, though it may temporarily rest and slide away from this base, requiring the powder to be dry and free-flowing.

The various possible arrangements are described by Marshall.[8] The outlet air will carry the remainder of the powder or 'fines', which need to be separated in an external cyclone, before discharge of this air to the atmosphere through a stack. A decision is required on the maximum amount of powder that can be emitted, determined by local or national legislative requirements. Knowledge of its powder particle size and inherent density of and the air flow available and the temperature at that point enables the cyclone(s) to be selected (say 99% efficiency at 20 μm). The fundamentals of powder separation are described in Perry,[2] with particular reference to the early work of Lapple and Shepherd.

2.8. Agglomeration

Agglomeration for instant coffee, that is, as granules formed from, or simultaneously with, spray-dried powder, has been used since about 1968. The granule sizes can average about 1400 μm, though with a range about the average, which again may be expressed on a probability linear chart from screen analysis. Granules are generally darker in colour than powders, and provide favourable comparison of appearance with roast and ground coffee. Jensen[30] has described the numerous methods available for agglomerating foodstuff powders in general, especially for milk powders (where the process is known as instantising, since conventional dusty skim-milk powders are not instant). He divides the processes into either 'rewet' methods, subdivided by use of steam condensation usually on pre-pulverised particles or of water droplets; or, 'straight-through' methods, to cause cohesion. Many foodstuff powders can be more readily agglomerated by addition of sucrose (e.g. cocoa powder); but instant coffee itself presents some special problems, agglomeration commonly being achieved by making use of thermoplasticity at higher surface moisture contents. A selection of patents describing techniques used is given in Table 9; in addition, there are patents involving the use of specialised agglomerator nozzles (with multi-annular orifices), which may or may not be suitable for instant coffees. After the initial wetting, the formed agglomerates need to be dried down to normal instant moisture content, for which a fluidised bed may be used.

Table 9
Typical agglomeration patents for instant coffee

Patent No.	Date	Assignee
USP 2,977,203	1961	General Foods Corporation
3,554,760	1971	
3,615,670	1971	
3,695,165	1973	
USP 3,514,300	1970	Nestlé
BP 1,176,320	1967	
USP 3,679,416	1972	Chock Full o' Nuts Corporation
USP 3,966,975	1974	Niro Atomizer A/S
BP 1,385,192	1974	
USP 3,615,669	1971	Procter & Gamble

3. PROCESS FACTORS IN FREEZE-DRYING

3.1. Methods

Freeze-drying is a relatively new technique for the food industry (post Second World War), and may be used for both liquid and solid foodstuffs. It was extensively studied and developed in the early 1950s and 60s, and numerous texts and papers describe its theory and practice with special reference to the food industry, primarily that concerned with the production of solid foodstuffs. See, for example, the reviews in references 31, 32 and 33, culminating in an extensive review by Judson King.[34] Although the freeze-drying of pre-frozen slabs of foodstuff liquids and extracts, requiring the subsequent break-up by grinding into dried particles (of irregular shape and size), has been used, it is now more general practice, following issuance of patents around 1965–67,[35] first to granulate frozen slabs of coffee extract (which may also be foamed by inert gas inclusion) into granules approximating in size to that required in the final product. Beds of these frozen granulates are loaded into and held in trays of special design during freeze-drying to facilitate the removal of moisture, primarily by sublimation of ice, in a reasonable time. The trays rest on heated shelves or are subjected to radiant heat, at controlled surface temperatures, within a chamber, while the actual temperature of the frozen material is kept sufficiently low to maintain this condition. The water vapour as it is formed has to be continuously removed by maintaining a pressure differential between that of the water in equilibrium with the ice and a lower pressure in the exit duct from the chamber. Although this can be done with large volumes of very dry air at atmospheric pressure, in practice a very high vacuum is applied and the exiting water vapour is condensed as ice (to be continually removed) on condensers, with surfaces kept at a temperature below that of the ice in the drying product by flowing refrigerant liquids. Freeze-drying is generally conducted in a batch manner, in chambers with internal/external condensers. Other, newer designs are semi-continuous and some are fully continuous with static beds and with agitated particles, despite the mechanical problems they present.

As with spray-drying, freeze-dried coffee physical characteristics (colour, bulk density, granule size and strength), and of course, their flavour quality, are important.[36]

Variants of freeze-drier design and operation have been discussed in a number of papers, apart from those already cited;[37-39] and some deal specifically with the freeze-drying of coffee extracts,[40-42] for which there are also many patented procedures.[35] There have also been a large number of fundamental studies in the field of combined heat/mass transfer, notably by King, Karel, Thijssen and their corresponding colleagues, to whom reference will be made in the following sections.

3.2. Mechanism of Water Removal

The water vapour pressure–temperature relationship for ice at equilibrium in the range of temperatures of operating interest for freeze-drying (as shown in Fig. 3) is important. Equally important is the phase diagram for ice–liquid–solids water content at different (low) temperatures in the material being freeze-dried; such a partial diagram for typical coffee extracts is shown in Fig. 4. The curves available will show some differences according to the blend and roast degree of the original coffee and its extraction yield (e.g. Gane's extract is of home-brewed coffee). The curves dip towards a temperature of about $-24°C$. Riedel[43] has provided enthalpy data over the whole range of extract concentrations down to $-60°C$ for an unspecified coffee extract. At $-24°C$, Riedel finds that with 77% coffee solids, there is still 23% water that is 'unfreezable' even at lower temperatures, equivalent to 0.30 kg per kilogram of solids (or starting at this concentration there is no ice formation on freezing). Such a phenomenon is familiar with many foodstuffs containing water, and macromolecular substances (proteins and carbohydrates) including sugars, and indeed with some inorganic solutions. The only water that can be said to be frozen is that which is 'free', whereas 'bound' water is inherently fixed in some measure at that temperature. This can lead to the

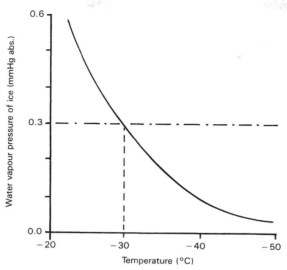

Fig. 3. Equilibrium relationship between the water vapour pressure of ice (mmHg absolute) and temperature (°C). –·–·–, Typical operating vacuum (pressure) in freeze-drying coffee extract.

identification of so-called eutectic mixtures within a complete phase diagram. The presence of eutectics is disputed in solidified coffee extracts, but it is evident that there is also a semi-solid matrix containing liquid water, through which the ice is distributed. This semi-solid matrix is considered to have a viscosity dependent upon temperature, 'water' content and the nature of the solids, as considered in the 'collapse' temperature concept of Flink and Karel.[44] Since liquid foodstuffs have no cellular structure, it is necessary to maintain some rigidity in the matrix so that it does not collapse with the gradually created pores in the sublimating ice during freeze-drying, which provide channels of exit for the water vapour to the outside. Otherwise 'puffing' will occur, together with general difficulty in and slow rates of drying. This difficulty is particularly experienced with liquid foodstuffs containing simple sugars (glucose and fructose) in large amounts (e.g. fruit juices), so that temperatures need to be brought down to as low as −54°C, primarily to ensure sufficient rigidity to the matrix. With coffee extracts on the other hand, −24°C would be generally regarded as sufficiently low to freeze all the water that can be frozen, and down to, say, −30°C to ensure rigidity, no doubt on account of the presence of polysaccharides and other high-molecular-weight substances in amounts dependent upon yield. 'Collapse'

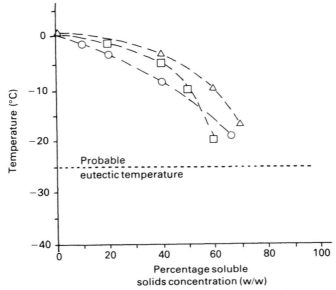

Fig. 4. Phase diagram for three coffee extracts, relating temperature (°C) of incipient freezing with percentage soluble solids concentration (w/w) of extract: ○--○, data of Gane; □--□, data of Sivetz and Desrosier[67]; △--△, data of Bomben et al.[13]

temperatures are extrapolatable to 'fusion' and 'sticky-point' temperatures in spray drying (see Volume 1, Chapter 2).

Supposing that a coffee extract is regarded as completely frozen at −24°C (water vapour pressure from ice, 0·55 mmHg absolute or 0·55 torr = 7·3 × 10^{-4} bar = 73 Pa), then a typical operating vacuum needed is 0·3 mm, corresponding to an ice temperature of −30°C. To maintain a flow of water vapour through various pressure drops away to the condensers requires a lower water pressure and temperature at the condenser of the order of 0·03 mm and −50°C respectively.

In order to convert ice to water vapour, latent heat must be supplied (of the order of 685 cal g^{-1}) to the so-called 'ice-front'. In practice this heat may be supplied from the heat source in a number of ways. With a frozen slab firmly and completely in contact with a heated surface, heat transfer will take place by conduction through the frozen layers to the outer surface of the slab, from which the water vapour will emerge, initially from the surface but no doubt later from within the drying material until all is dried. Fortunately, the thermal conductivity of ice is fairly high (1.9×10^{-2} kcal cm^{-1} h^{-1} °C^{-1}; or 2.2 Wm^{-1} k^{-1}) but increases with decreasing temperature also for the frozen layer, so that the process becomes 'mass-transfer controlled', i.e. the major resistance to drying rate is that of moisture removal. If the contact is irregular, then the water vapour will sublime and escape from that part of the material closest to the heater plate, requiring that heat then be transferred through the dried layer as drying proceeds. This will also be so in the freeze-drying of frozen granulates; and in other instances the heat transfer may be a mixture of both. The thermal conductivity of dried material (which is a combination of that of the dried material itself and of entrained gases/vapours) is not, however, good (about 1/100 that of ice); on the other hand, mass transfer, especially with granules, is favourable. The process is then said to be heat-transfer controlled. An alternative and often preferred method of heat application is by radiant heat, which inevitably provides for conduction of heat through the already dried layers.

The typical temperature profiles that result, both in the frozen and porous layers, are shown in Fig. 5, as given by Thijssen,[12] together with typical water concentration profiles. The actual ice-front temperature is

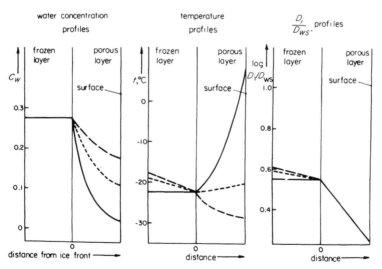

Fig. 5. Profiles of water concentration, temperature and selective diffusivity in freeze drying. Heat is applied through the frozen layer, ———; the porous layer, — or both, – –. From Bomben et al.[13] Reproduced by permission of Academic Press, San Diego, CA.

determined during drying by the interaction of heat-transfer and mass-transfer effect, i.e. by the surface temperature, chamber vapour pressure and the dry layer properties of thermal conductivity and permeability, and will be generally constant in heat-transfer-controlled drying. With mass-transfer-controlled drying (heat through the frozen layer), the ice-front temperature and pressure become also dependent upon the time of drying, since the vapour pressure of the ice depends upon the thickness of the dried layer. Karel[45] has presented equations for the time of drying of a frozen slab for both types of heat transfer. For the case of heat transfer through the dried layer he gives, in engineering units:

$$t = \frac{L^2 \rho (m_1 - m_f)}{8b(p_1 - p_s)}$$

where L is the thickness of the slab (ft) or layer of granules; m_1 and m_f are the initial and final moisture contents, pounds per pound of solids; ρ is the density of the solids in the slab (lb ft^{-3}); b is the permeability of the dried layer (lb ft^{-1} h^{-1} torr^{-1}); and p_1 and p_s are the vapour pressures. There is an equivalent equation based on heat transfer:

$$t = \frac{L^2 \rho (m_1 - m_f) \Delta H_s}{8 k_d (T_s - T_i)}$$

where ΔH_s is the latent heat of fusion (Btu per lb); T_s and T_i are the surface and ice temperatures; and k_d is the thermal conductivity, all in consistent units.

Such equations are useful in determining the effect of various variables on the time of drying; for a given initial concentration and required final concentration of soluble solids, the time of drying will depend inversely on the difference between the surface temperature and the ice temperature. In practice, a maximum permissible surface temperature is specified (say 40–50°C, for reasons of potential thermal degradation). At first, a fixed platen temperature allows the surface to reach this temperature (at a rate which can be calculated), after which it is gradually lowered (often stepwise); the ice-front temperature must be below the critical value. There is a third period in which the ice has disappeared, when the heat will raise the dry layer temperature; though only little discussed by various authors, desorptive drying of the remaining bound water has to take place to obtain the final moisture content desired. It is apparent that the release of this moisture will not take place until the higher temperatures are reached.

For the purpose of examining the effect of process variables both for water removal and retention of volatile compounds, Thijssen and Rulkens[46] use a model for drying based on the so-called uniformly retreating ice front. In the initial freezing of a slab, the ice is assumed to be present in the form of equally spaced tubular crystals (from unidirectional freezing) perpendicular to the surface and extending through the matrix. The volatile compounds will necessarily remain in the matrix. As sublimation of ice proceeds, so pores develop to provide escape for water vapour to the free atmosphere, and from the water in the matrix itself. The

Table 10
Influence on drying rate in freeze-drying of various parameters[13,49]

Increasing value of variable	Effect on drying rate with heat applied through:	
	Dried layers	Frozen layers
Pore size, from slow freezing rate	Increase	Increase (more pronounced)
Soluble solids concentration (initial)	Decrease	Decrease
	(NB overall product rate is markedly increased)	
Thickness of slab (also layer of granules)	Decrease	Decrease
	(NB there may be a slight increase in overall product rate)	
Granule size	—	—
Ice-front temperature Surface temperature Plate temperature Chamber pressure (or at drying surface)	Linked variables (see Karel and Thijssen equations). In general, increase with increase of difference between relevant temperature and pressures	

porosity of the dry layer will be equal to the volume fraction of ice in the material at the sublimation temperature. The mean pore diameter, however, importantly increases with decreasing rate in the initial freezing (as shown also by others[47,48]), and can be determined and checked by permeability measurements for nitrogen. Thijssen *et al.* provide differential equations for the water removal and derive a number of conclusions on the rate of drying as affected by changes in variables under the two systems of control, as indicated in Table 10.

Oetjen and Ellenberg[38] especially, and also Orlin,[42] studied the high drying rates achieved with systems of agitated particles with various rates of mixing.

3.3. Mechanism of Volatile Compound Retention

The important aspect of degree of volatile compound retention has been studied and described by Thijssen, Karel, Judson King and their respective colleagues. It has been known for some time that the retention is generally good in freeze-drying; however, it is now evident that various variables in the pre-freezing and in the freeze-drying itself markedly influence the degree of retention actually obtained. Small laboratory freeze driers are available for experimental work, and numerous experimental studies have been reported with different volatile compounds and concentrations, foodstuff solutions and drying conditions, though the results are sometimes rather conflicting.

Thijssen and co-workers[12,13,20] have applied the selective diffusion concept, already discussed in section 2.2, to an examination of the variables in freeze-drying, based upon the model and equations already described in section 3.2 together with experimental studies.[49] Changes in the ratio D_j/D_w are considered within the matrix, together with a more formal mathematical analysis of aroma transport and that of water. The freezing rate was shown to have a marked effect on retention, and should be kept as low as possible. The retention, under identical conditions, differs according to the liquid diffusivity of the volatile component, being greater the lower the numerical value, and is not related to boiling point or relative volatility. Low diffusion coefficients are indicated by high molecular weights (or molal volumes). The molecular weight of the soluble solids, which can vary in coffee extracts, is also influential. Some effects are indicated in Table 11.

The original concept by Rey and Bastien[50] was that volatile compounds were retained by surface adsorption on the dry layers; but this is now discounted following further experimental work[45] and an entrapment mechanism is favoured. However, Flink and Karel[44,51] differ in the nature of this entrapment by postulating the existence of micro-regions, the size of which depends, together with other factors, on the solubility of the volatile compound in the aqueous solution.

Further experimental work of related interest[45] is that in which already freeze-dried products are humidified, when it was found that only when the moisture content was above the monolayer level (from the Brunauer–Emmett–Teller equation) did release of volatile compounds take place.

Later experimental work on the retention of volatile compounds is described in references 52 and 53. In the latter paper, work was carried out with foamed coffee extracts (at 27% w/w soluble solids) using commercial plant (Atlas) in a number of different ways (slow and fast freezing, and different freeze-drying conditions). Differences of retention

Table 11
Influence on volatile compound retention in freeze-drying of various parameters[13,49]

Increasing value of variable	Effect on volatile compound retention with heat applied through:	
	Dried layers	Frozen layers
Freezing rate	*Decrease*	*Decrease*
	Effect greatest for high ice-front temperatures, low initial soluble-solids concentrations and low drying rates	
Drying rate generally	*Increase* for a constant ice-front temperature. Slight *decrease* with increase of plate temperature (and ice-front temperature)	
Molecular size of volatile compound	*Increase* by effect on D/D_w value for given water concentrations (NB effect may differ according to concentration of volatile compound)	
Soluble solids concentration (initial)	Probably small overall effect	
Thickness of slab	*Decrease*, by decrease in drying rate	
Granule size (diameter)	Optimum diameter. At constant load, *increase* with particle diameter when small, *decrease* when large	
Chamber pressure	Linked variables. Increased chamber pressure at constant ice-front temperature *decrease*. Decreased ice-front temperature *increase* (provided temperature decrease not accompanied by decrease in drying rate, on heating through dry layer). *Increase* with decreasing plate temperature (for constant ice temperature and chamber pressure). At constant plate temperature, *decrease* with increasing ice-front temperature	
Plate temperature		
Surface temperature		
Ice-front temperature		

were found in particular groups of volatile compounds (assessed as peaks in GC analysis), but overall significant flavour differences (by panel testing and statistical analysis) could not be found. Voilley and Simatos[54] have carried out experimental work on the retention of volatile compounds with particular reference to the molecular weight of the soluble solids. No work appears to have been reported on the freeze-drying of agitated particles, where the high drying rates achieved might be expected to favour volatile compound retention.

4. PROCESS FACTORS IN PRE-CONCENTRATION

4.1. Evaporation

The use of evaporation to strip out volatile compounds from coffee extract has already been discussed in Chapter 5. According to the level of stripping required, additional evaporation may be needed for bulk evaporation to achieve a required soluble-solids concentration. Both stripping and bulk evaporation may be carried out by the same kind of evaporating method. For evaporating foodstuff liquids, including coffee extracts, film methods are now preferred, with their relatively short residence times (i.e. of the order of seconds) and therefore reduced risks of thermal degradation.

These film evaporators were the subject of a symposium in London, to which reference should be made.[55-58] Variants of operation, e.g. multi-stage operation with either forward or back flow, are dealt with in numerous texts, e.g. reference 2. Certain characteristics of coffee extracts have to be considered, in particular their capacity for foaming and the marked increase in viscosity that occurs after a concentration of about 42% w/w soluble solids has been reached (see Volume 1, Chapter 2); concentrations above this level become increasingly difficult to reach in practice, though the wiped film evaporator is claimed to reach about 70% w/w. Choice of surface temperatures is also important, in minimising fouling of heat exchange surfaces through deposition and in minimising taste changes in the extract. No particular information has been published on desirable residence times and temperatures, nor upon the nature of deposits, which are likely to be proteinaceous (as with milk).

4.2. Freeze-concentration

Freeze-concentration is a low-temperature method of removing water from an aqueous solution of solids, and has many attractions for the food industry including the coffee industry. In examining the phase diagram for coffee extracts (Fig. 4), it can be seen that a solution of, say, 25% w/w coffee solubles (i.e. 25% solids: 75% water) will begin to freeze at about $-2°C$; a further lowering of the temperature increases the amount of ice formed, till at $-4°C$ the concentration of the soluble solids in the remaining liquid will be about 40% w/w. At that temperature the solution is in equilibrium, with ice constituting 50% of the original weight of water, i.e. no more ice is formed—unless of course the temperature is lowered still further. It is now necessary to separate and remove this ice from the solution, with as little adhering solution as possible; otherwise there will be unwanted loss of soluble solids as a significant proportion of that in the original solution or extract. The formation of this ice in the most physically desirable condition (i.e. as regards crystal size and shape) is closely connected with its efficient subsequent separation. In general, larger ice crystals are preferred in that they are associated with lower amounts of solution and are more easily removable.

Consequently, a considerable number of different designs are available in commercial plant, many of which are also patented, and are also available for non-food uses (e.g. desalination of sea water to obtain pure water). Thijssen[59] subdivides these designs into crystallisers which are indirectly cooled (i.e. by non-evaporative cooling) either internally or externally; the former are further subdivided into those that produce either solidified or 'pumpable' suspensions, and the latter are subdivided into three subgroups. Separation is achieved by one of three methods, hydraulic piston or screw presses to press out residual liquid, filtering centrifugals, and, more recently, wash columns. Filtering centrifugals enable the washing of the cake ice either with fresh water, or with melted ice, to the limit of acceptable dilution of soluble solids; but can have the disadvantage of the loss of volatile compounds during the spinning in contact with an excess of air. The loss of volatile compounds is otherwise minimal and constitutes a marked advantage of freeze-concentration over evaporation. The lower latent heat of fusion of ice compared with that of the vaporisation of water constitutes another 'theoretical' advantage. The general operation of all these designs is described by Thijssen,[12,59] who also in these texts discusses the fundamental principles of ice crystallisation and separation behind the designs.

The level of concentration of coffee extracts that can be achieved by this technique is, like evaporation, limited by viscosity factors, so that in practice about 40% w/w is possible. Some coffee extracts on cooling may deposit (e.g. in 30 min at $0°C$ for maximum precipitation of a few per cent) a sludge, which needs to be removed otherwise it can interfere with effective freeze-concentration (e.g. British Patent 1,129,975, General Foods Corporation, 1967).

Much study has also been directed towards the relative economics of freeze-concentration and multi-stage evaporation, for example by Thijssen[12] and more recently by Van Pelt.[60,61] In the latter paper, the economics of combined pre-concentration and drying (both spray- and freeze-drying) for instant coffee manufacture are considered, together with the concepts of the use of freeze-concentration on the first part of the draw from an extractor and of evaporation on the second part. It is of interest that, for the same weight of water removed, the energy consumption for freeze-concentration is calculated to be comparable to that of single-effect evaporation but not to multi-effect; capital costs are some 2·5 times those for single-stage evaporation (costing in 1977).

4.3. Reverse Osmosis

The technique (together with other forms of membrane concentration) for the removal of water is probably much less used than the foregoing with coffee extracts, though it could have advantages in handling dilute process streams. The principles are described in a number of texts,[13,62-65] though none of these has particular reference to coffee extracts. In order to retain a high proportion of volatile compounds, especially those of low molecular weight, it is generally considered that 'tight' membranes are required with some reduction in water permeability rates, though again there is the question as to which coffee volatile compounds really require to be retained.

5. PROCESS EQUIPMENT

It is probably fair to say that most of the equipment used for carrying out the spray-drying/agglomeration processes described in this chapter will be designed by the major instant coffee manufacturers themselves, though specific items of plant and whole installations are available from a number of plant manufacturers. Sivetz and Foote[66] and Sivetz and Desrosier[67] offer details of much of the plant and of practical modes of operation including start-ups. Various descriptions of the total instant coffee process, which will differ according to location, are available in trade journals and elsewhere. Many accounts must, however, be read with caution.

5.1. Spray Driers

The broad outlines of design requirements have been given in section 2. In particular, Niro Atomizer A/S, of Copenhagen, Denmark, supply driers of various capacities for spray-drying coffee extract (Fig. 6); however, there are many other manufacturers of spray driers throughout the world. With centrifugal pressure nozzles, a feed pump with a capacity and end-pressure capable of handling the necessary nozzle pressures is required. Multi-stroke reciprocating pumps will provide the necessary

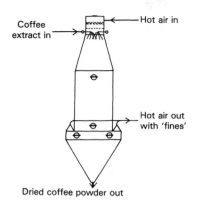

Fig. 6. Typical spray drier for coffee extract, with twin centrifugal pressure nozzle assembly and a primary separation at base of tower. By courtesy of Niro Atomizer A/S, Copenhagen.

non-pulsating flow. The hot air is generally provided by a furnace burning high-quality fuel oil (low sulphur content) or natural gas with a primary source of air, admixing with secondary air, for direct contact with the spray. It is convenient to have a control to the furnace from a fixed outlet temperature where capacity is being varied. There are numerous manufacturers of cyclones to be externally situated (to the spray drier) and of multiclones for the powder–air separation, who provide characteristic curves for their performance in terms of efficiency at different particle size and density (though generally of solid particles which may not be the case with instant coffee). The collection of powder and its subsequent storage prior to packing are conveniently handled in large installations by tote systems. The fines separated may cause logistic problems, though various solutions are to be found. These include refeeding back close to the nozzle of the drier, or redissolving in coffee extract/water for reprocessing or as part of the feed for subsequent agglomeration.

The Birs Technica spray-drying towers[68] are of interest in that they employ only warm air in the countercurrent movement to a spray within a tower some 200 ft (61 m) high. However, there is no evidence of their successful use with coffee extracts, especially in respect of volatile compound retention.

5.2. Agglomerators

Two plant manufacturers describe the construction of agglomerators suitable for instant coffee, Niro,[30] and Lurgi of Munich. Both of these designs are based upon rotating discs and use steam/water as the wetting medium followed by bed drying.

5.3. Evaporators

Evaporation is often associated with pre-evaporative stripping (as described in Chapter 5). The film-type evaporators are preferred in each case, and there are many manufacturers. In particular, there is the APV Co. Plate Evaporator, first patented in its present form in 1958 (the genesis of which was described by Dummett[69] and its construction and operation by Gray[56]); the centrifugal evaporator with its very short residence time (Centritherm), of Alfa Laval;[57] the long tubular evaporator, also with rising/falling films, of Wiegand;[58] the wiped film with either fixed or flexible blades, of Luwa and Rodney Hunt in the USA; the Stratavap, of the Balfour Group in Leven, Scotland; and the horizontal film of Burnett and Rolfe, Rochester, England.

Modern steam economy demands the use of either multi-effect evaporation and/or thermo-recompression, though three effects, unlike the sugar industry, would be considered the maximum suitable for coffee extracts in bulk evaporation.

5.4. Freeze Concentrators

The wide variety of designs available have been discussed in section 4.2. Van Pelt[60] has described the operation of the Grenco freeze concentrator with its wash column, and Orlin[42] the Krauss-Maffei concentrator as used for coffee extract. Thijssen[59] lists a number of other commercial designs: Gasquet and Daubron (using filter presses for the final ice separation) from France; Union Carbide, Struthers and Phillips from the USA; and Krause-Linde from Germany.

5.5. Freeze Driers

There have been a number of published papers on freeze-drier installations suitable for coffee extracts and on their operation, notably by Sivetz and Desrosier[67] who describe the various commercial types available. The main suppliers are FMC and Stokes in the USA; Leybold, described by Oetjen,[37] and Lurgi, described by Orlin[42] in Germany; and Atlas (Conrad driers) of Copenhagen, Denmark, described by Lorentzen.[39,70] In addition, Poulsen[71] has discussed the economics of coffee freeze-drier installations.

These are divisible into batch chamber driers (e.g. FMC, Stokes and smaller Conrad models), continuous intermittent driers (e.g. Leybold and Conrad), and more truly continuous with agitated particles (e.g. Lurgi). Whereas the first two types use specially designed trays to hold the frozen coffee granules, platen heating is generally adopted except in the Conrad models which use radiant heat. There are other basic engineering differences, especially in the design and situation of the ice condensers.

A freeze-drier installation for coffee extract is, however, much more than the drier itself. Plant is required for the initial freezing of the coffee extract to around −40°C (whether foamed by inert gas injection or not), which can be carried out on so-called freezing belts (e.g. those designed by Sandvik of Sweden) or on freezing drums. The frozen slabs need to be granulated in the frozen condition and loaded into trays. The refrigeration system (liquid ammonia at −45°C is usually satisfactory) and vacuum system (steam ejectors rather than reciprocating or rotary pumps are favoured) are also major items of plant. Tray handling of frozen and finished products are also important logistic elements of the installation. Capacities of the order of 1 ton of product per 24 h are feasible.

5.6. Other Dryers

Continuous vacuum belt dryers (that is, operating at a lower vacuum than in freeze-drying) are used for a number of liquid foodstuffs, and there is some evidence of their use for coffee extracts. In conjunction with chilled drums,[67] instant coffee flakes can be made, e.g. US Patent 3,652,293 (Procter & Gamble, 1972). An interesting variant of the continuous vacuum belt dryer is the use of microwave heating described by Meisel[72] on liquid foodstuffs introduced into the drier in paste and foamed form, with a drying time of some 25–40 min.

A so-called combination dryer has been described for use with foodstuff liquids.[25,73] This combines a typical spray drier with moving bed drying to give two separately controllable stages of drying with claimed advantage in the retention of volatile compounds.

5.7. Process Control

The commercial manufacture of instant coffee, whether as powder or as granules (agglomerated or freeze-dried), requires close quality monitoring of at least the following physical properties. Bulk density is determined either under specified free-flow conditions or when packed under a specified vibration (to simulate actual machine packing). There is a draft International Standard (DISO 8460–1985), with recommended procedures. Moisture content (see Volume 1, Chapter 2) is generally measured by a vacuum oven method, e.g. IS 3726–1983, or by the Karl

Fischer titration method. Colour is measured by one of the reflectance meters available, as with roasted coffee, though in this case colour is measured directly on the sample 'as is' placed in a shallow dish, i.e. without any screening or pre-treatment. Screen analyses, e.g. according to ISO 7532–1985, are carried out for instant coffee with specific reference to the quantity of fines produced, at whatever screen size is judged to pass 'fines'. Screen analysis data, when plotted as aperture size (linear) vs. percentage cumulative amount at a given size on a probability scale, usually provide a linear plot (except at the ends), so that the screen size at 50% cumulative amount gives a reasonable estimate of average particle size. Screen analyses carried out under different sets of conditions are often used as a means of assessing the hardness of granules as described by Peterson for freeze-dried[36] and by Sylla and Schrein[74] for agglomerates. ISO 7534–1985 describes a method for determining insoluble matter present in traces (below 1000 ppm) at a size more than 100 μm; sediment pad tests (as used in the milk industry) are often carried out to monitor freedom from fine inert coffee grounds. As the flowability of powders is of concern, empirical quantitative tests are available; however, there is usually little problem with granules. Experience usually determines the exact settings of plant process variables to achieve correct specifications. Quality control in respect of flavour will be carried out by trained operators and checked by panels; now generally, GC determinations on finished and intermediate products will be used as back-up to ensure satisfactory flavour retention.

With liquid extract intermediates, the concentration of soluble solids is an important parameter to be determined, methods for which are described in Volume 1, Chapter 2.

5.8. Dust and Fire Hazards

This subject is of practical importance in the design and operation of spray driers and associated powder handling equipment. Spontaneous self-heating and ignition potentials for instant coffee and instant chicory powders have been discussed by Raemy and Lambelet[75] following their calorimetric curve analyses of these substances. In their paper exothermic heat data and, though not practically applicable, ignition temperatures are given (also for spent coffee grounds, which has a particularly high heat content corresponding to its high carbohydrate content). This subject in all its practical aspects is well dealt with, exampling the spray-drying of skim milk powder, in a recent paper.[76]

REFERENCES

1. King, F. J., *Proc. 9th Coll. ASIC*, 1980, 237–50.
2. Perry, J. H., *Chemical Engineers' Handbook*, 4th Edn, McGraw-Hill, New York, 1976.
3. Masters, K., *Spray Drying*, 2nd Edn, George Godwin, London, 1976.
4. Seltzer, E. and Settelmeyer, J. T., in *Advances in Food Research*, Vol. 2, Eds E. M. Mrak and G. F. Stewart, Academic Press, New York, 1949, 399–520.
5. Clarke, R. J., Spray drying of liquid foodstuffs, *1st International Congress of Food Science and Technology, London, 1963*, Vol. 1, Gordon & Breach, London, 1963, 3–12.
6. Thijssen, H. A. C. and Kerkhof, P. J. A. M., *J. Fd Proc. Engng*, 1977, **1**, 129–47.
7. Thijssen, H. A. C., *Lebensm. Wiss. Technol.*, 1979, **12**, 308–17.
8. Marshall, W. R., *Atomization and Spray Drying*, 1954 Chem. Engng Prog. Monog. Series, Vol. 50, American Institute of Chemical Engineers, New York.
9. Dombrowski, N. and Munday, G., in *Biochemical and Biological Engineering Science*, Vol. 2, Academic Press, New York, 1968, 209–320.
10. Clarke, R. J., *Proc. Biochem.*, 1969, **4**(2), 37–42.
11. Dombrowski, N. and Wolfsohn, D. L., *Trans. Inst. Chem. Engrs*, 1972, **50**, 259–69.
12. Thijssen, H. A. C., *Proc. 4th Coll. ASIC*, 1969, 108–17.
13. Bomben, J. L., Bruin, S., Thijssen, H. A. C. and Merson, R. L., in *Advances in Food Research*, Vol. 20, Eds C. O. Chichester, E. M. Mrak and G. F. Stewart, Academic Press, New York, 1973, 2–111.
14. Sjenitzer, F., *Chem. Engng Sci.*, 1952, **1**, 101–17.
15. Rulkens, W. H. and Thijssen, H. A. C., *Trans. Inst. Chem. Engrs (UK)*, 1969, **47**, T292–8.
16. Menting, L. C., PhD thesis, University of Eindhoven, 1969.
17. Menting, L. C., Hoogstad, B. and Thijssen, H. A. C., *J. Fd Technol.*, 1970, **5**, 111–26.
18. Kerkhof, P. J. A. M. and Schroeber, W. J. A. H., in *Advances in Preconcentration and Dehydration of Foods*, Ed. A. Spicer, Applied Science Publishers, London, 1974, 349–98.
19. Thijssen, H. A. C., *Proc. 4th Coll. ASIC*, 1969, 108–17.
20. Thijssen, H. A. C., *Proc. 6th Coll. ASIC*, 1973, 222–31.
21. Menting, L. C., Hoogstad, B. and Thijssen, H. A. C., *J. Fd Technol.*, 1970, **5**, 127–39.
22. Zakerian, J. A. and King, C. J., *Ind. Engng Chem. Process Res. Dev.*, 1982, **81**, 107–13.
23. Kerkhof, P. J. A. M., *Proc. 8th Coll. ASIC*, 1977, 235–48.
24. Crank, J., *Mathematics of Diffusion*, Oxford University Press, Oxford, 1956.
25. Kerkhof, P. J. A. M. and Thijssen, H. A. C., *Amer. Inst. Chem. Engrs*, 1975, Symposium Series **73**(163), 33.
26. Kerkhof, P. J. A. M. and Thijssen, H. A. C., *Proc. Int. Symp. Aroma Research (Zeist)*, Wageningen, 1975, 167–92.

27. Van Lijn, J., Rulkens, W. H. and Kerkhof, P. J. A. M., *Proc. Int. Symp. on Heat and Mass Transfer in Food Engineering*, Dechema, Frankfurt, 1972, Paper F3, 1–23.
28. Menting, L. C. and Hoogstad, B., *J. Fd Sci.*, 1967, **32**, 87–90.
29. Rulkens, W. H. and Thijssen, H. A. C., *J. Fd Technol.*, 1972, **7**, 95–105.
30. Jensen, J. D., *Food Technol.*, 1975, June, 60–71.
31. Ray, L., *Lyophilization*, Hermann, Paris, 1964.
32. Burke, R. F. and Decareau, R. V., in *Advances in Food Research*, Vol. 13, Academic Press, New York, 1964, 1–88.
33. Meffert, H. F., in *Dechema Monographien*, No. 1125–43, **63**, Dechema, Frankfurt, 1969, 127–51.
34. King, C. J., *Freeze Drying of Foods, Critical Reviews in Food Technology*, Vol. 1, Chemical Rubber Publ. Co., West Palm Beach, Fla., 1971, 379.
35. Pintauro, N., in *Coffee Solubilization*, Noyes Data Corporation, Park Ridge, NJ, 1975, 126–36.
36. Peterson, E. E., *Tea and Coffee Trade J.*, 1971 (Oct.), 16–20 and 46.
37. Oetjen, G. W., *Proc. 18th Italian National Congress on Refrigeration*, Padua, 1969.
38. Oetjen, G. W. and Ellenberg, H. J., *Proc. XIIIth Int. Congress of the Institute of Refrigeration (Commission X)*, Lausanne, 1969.
39. Lorentzen, H., *Chem. Ind.*, 1979, 165–9.
40. Castiglioni, N. V., *Industrie Alimentari*, 1966, 62–5.
41. Picallo, J. A. C., *Proc. XIIth Int. Congress of the Institute of Refrigeration*, Madrid, 1967, 19, Paper 6c.
42. Orlin, P., *Proc. 6th Coll. ASIC*, 1973, 240–54.
43. Riedel, L., *Chem. Mikrobiol. Technol. Lebensm.*, 1974, **4**, 108–12.
44. Flink, J. and Karel, M., *J. Agric. Fd Chem.*, 1970, **18**, 295–7.
45. Karel, M., in *Advances in Preconcentration and Dehydration of Foods*, Ed. A. Spicer, Applied Science Publishers, London, 1974, 45–94.
46. Thijssen, H. A. C. and Rulkens, W. H., in *Proc. XIIIth Int. Congress of the Institute of Refrigeration (Commission X)*, Lausanne, 1969.
47. King, C. J., Lem, W. H. and Sandall, Q. C., *Food Technol.*, 1968, **22**, 100, 1302.
48. Kramers, H., in *Fundamental Aspects of the Dehydration of Foods*, SCI, London, 1958.
49. Rulkens, W. H. and Thijssen, H. A. C., *J. Food Technol. (UK)*, 1972, **7**, 79–93.
50. Rey, L. and Bastien, M. C., in *Freeze Drying of Foods*, Ed. F. R. Fisher, NRC, Washington, 1962, 25.
51. Flink, J. and Karel, M., *J. Fd Technol. (UK)*, 1972.
52. Peterson, E. E., Lorentzen, J. and Flink, J., *J. Fd Sci.*, 1973, **38**, 119.
53. Fosbol, P. and Peterson, E. E., *Proc 9th Coll. ASIC*, 1980, 251–6.
54. Voilley, A. and Simatos, D., in *Food Engineering Processes*, Vol. 1, Eds P. Linko *et al.*, Applied Science Publishers, London, 1980.
55. Clarke, R. J., *J. Appl. Chem. Biotechnol.*, 1971, **21**, 349–50.
56. Gray, R. M., *J. Appl. Chem. Biotechnol.*, 1971, **21**, 359–62.
57. Shinn, B. E., *J. Appl. Chem. Biotechnol.*, 1971, **21**, 366–71.
58. Wiegand, J., *J. Appl. Chem. Biotechnol.*, 1971, **21**, 351–8.

59. Thijssen, H. A. C., in *Dechema Monographien*, No. 1125–43, **63**, Dechema, Frankfurt, 1969, 153–77.
60. Van Pelt, W. H. J. M., *Proc. 6th Coll. ASIC*, 1973, 155–62.
61. Van Pelt, W. H. J. M., *Proc. 8th Coll. ASIC*, 1977, 211–16.
62. Merson, R. L., Ginnette, L. F. and Morgan, A. I., in *Dechema Monographien*, No. 1125–43, **63**, Dechema, Frankfurt, 1969, 179–202.
63. Madsen, R. F., in *Advances in Preconcentration and Dehydration of Foods*, Ed. A. Spicer, Applied Science Publishers, London, 1974, 251–302.
64. Michaels, A. S., in *Advances in Preconcentration and Dehydration of Foods*, Ed. A. Spicer, Applied Science Publishers, London, 1974, 213–50.
65. Pepper, D., *Proc. 3rd Int. Congress on Food Engineering*, Dublin, Applied Science Publishers, London, 1983, 621–8.
66. Sivetz, M. and Foote, M., *Coffee Technology*, Vol. 2, AVI, Westport, Conn., 1963.
67. Sivetz, M. and Desrosier, N. W., *Coffee Technology*, AVI, Westport, Conn., 1979.
68. Schultz, H. E., *Milchwissenschaft*, 1958, **13**, 393.
69. Dummett, G. A., *Chem. Ind.*, 1978, 17 June, 396–406.
70. Lorentzen, J., in *Advances in Preconcentration and Dehydration of Foods*, Ed. A. Spicer, Applied Science Publishers, London, 1974, 413–33.
71. Poulsen, V., *Proc. 6th Coll. ASIC*, 1973, 255–9.
72. Meisel, N., in *Advances in Preconcentration and Dehydration of Foods*, Ed. A. Spicer, Applied Science Publishers, London, 1974, 505–9.
73. Meade, R. E., *Food Technol.*, 1973, December, 18–20.
74. Sylla, K. F. and Schrein, J., *Proc. 7th Coll. ASIC*, 1975, 341–8.
75. Raemy, A. and Lambelet, P., *J. Fd Technol (UK)*, 1982, 451–60.
76. Beaver, P. F., *J. Fd Technol. (UK)*, 1985, 637–46.

Chapter 7

Packing of Roast and Instant Coffee

R. J. CLARKE*
Formerly of General Foods Ltd, Banbury, Oxon, UK

1. INTRODUCTION

The roast and instant coffee products of the previous chapters are not completely finished until they have been packed in suitable containers for sale to the consumer. With many foodstuffs this operation is a relatively simple matter; but roast coffee presents a number of problems which have had to be solved.

The central problem for roast coffee, especially roast and ground coffee, as already mentioned is the steady evolution of carbon dioxide gas (at a faster rate in the latter) in relatively large amounts per unit weight of coffee. Immediate packing of roast coffee would lead to a build-up of gas within a closed container, and possible explosion unless it was very strong or flexible. This build-up could also be potentially dangerous on opening. The use of an open or permeable package, except for short periods, is not acceptable on account of the marked effect of oxygen in the air on the stability and therefore shelf life of coffee, especially with moisture pick-up. These twin problems have been solved for roasted coffee in a variety of ways, which will be described in the following sections.

Instant coffee does not have this problem of the release of carbon dioxide; though volatile compound retention is important. It is also susceptible to the action of oxygen in the presence of water. Furthermore, being a hygroscopic substance, moisture pick-up from air ingress leads

---

* Present address: Ashby Cottage, Donnington, Chichester, Sussex, UK.

to 'caking' at around 7% w/w moisture content. The packing of instant coffee therefore also requires attention.

The problems of packaging roast coffee and their solution have been studied in detail by Heiss, Radtke and their colleagues at the Fraunhöfer Food and Packaging Institute of the Munich Technological University in a series of papers in German, e.g. references 1 to 5, and one in English (see reference 6).

2. PACKING OF ROAST WHOLE BEAN COFFEE

The scale of commercial or large-scale packing of roasted whole coffee beans is now relatively small in the industrialised countries compared with that of roast and ground. Such roast coffee is sold from speciality local coffee shops where in-house roasting is practised, and simple bags are used. Nevertheless, roasted whole beans are still sold from the larger manufacturers, as in France and in Spain, where, indeed, until recently the sale of roasted and ground coffee was prohibited.

2.1. Carbon Dioxide Evolution

Whether or not the roasted coffee is ground, it releases gases which consist mainly of carbon dioxide, CO_2 about 87%. The only difference is that on grinding roast coffee, there is a marked release of CO_2 during grinding itself, and the subsequent evolution of gas is at a faster rate due to the state of subdivision of the particles.

Roasted whole beans as a result of roasting contain entrapped gases, as described in Chapter 4, section 2.2. The quantity of CO_2 entrapped in roast whole beans is of the order of 2–5 ml (measured at NTP) per gram of roasted coffee, and probably higher[7] depending upon the method of measurement. Radtke[4] describes simple equipment for its measurement, which consists of measuring the pressure developed in a closed system with a manometer, and converting the readings to volume of gases (considered as CO_2 at NTP) since $PV =$ constant, at different times. Experimental times as long as 2500 h are required to reach equilibrium values. Radtke's interest lies primarily in determining pressure development due to CO_2 within closed packages. Barbara[7] measured CO_2 evolution from roasted whole beans to the air by a displacement method with hot water; flushing methods with, say, nitrogen gas and collection/absorption in alkaline solutions at atmospheric pressure by gravimetric determination can also be used. The quantity of CO_2 originally present, which is equal to that released after infinite time, is known to depend upon blend and roast level, i.e. the darker the roast, the higher the amount, and roasted robusta contain somewhat more than arabica. As there is continuous release of CO_2, which is a continuation of that released during roasting, the point of time zero is important in experimental determinations of content. Time zero should strictly be taken as immediately after roasting/cooling; though in packing studies that immediately on filling into packages will be of more interest.

Radtke[4] presents graphical data for two samples of roast whole beans and their decaffeinated counterparts, relating the amount of CO_2 released (in her laboratory equipment), until nearly asymptotic figures were reached. Table 1 gives a selection of these results, taken from the graphs.

The apparently greater and faster evolution from decaffeinated coffees is evident. These figures cannot, however, be taken as exactly corresponding to any initial content of CO_2 in the beans, as complete release of CO_2 would be impeded by the pressure (of CO_2) development in the closed system (reaching some 600 mmHg), nor are the figures fully asymptotic. Nevertheless, they give a good representation of the order of the amount and rate of release.

Little has been published on the mechanism of this release; but the diffusion flow of gas will be taking place under the influence of a concentration gradient to the outside atmosphere at negligible CO_2 content. The flow will also be hydrodynamic through pores and capillaries of the roasted coffee beans under the influence of a pressure differential

Table 1
Release of carbon dioxide[4] from roast whole bean coffee[a]

Time[b] (h)	Amount of CO_2 released (ml at NTP) per 100 g			
	Coffee 1	Coffee 2	Coffee 1 Decaff.	Coffee 2 Decaff.
0	0	0	0	0
200	180	200	275	220
400	250	265	340	275
1400	—	—	—	405[c]
1800	—	400[c]	450[c]	—
2500	378[c]	—	—	—

[a] Unspecified roast degree and blend.
[b] Time from start of measurement; time from roasting not given.
[c] Figures still slightly rising on curve.

of the CO_2 gas in the bean and outside. Both of these mechanisms imply an asymptotic approach to a maximum figure, which is found in practice.

Mathematically[8] the value of the ratio E (remaining amount of CO_2 at any time, t)/(initial amount of CO_2 in the roast coffee) will be related to the time, t, and a rate coefficient, k. The remaining amount of CO_2 is readily obtained by subtracting the released amount of CO_2 from the initial amount (which is the same as the amount released after infinite time). As the initial amount is constant, the remaining amount may be used as a parameter.

Fick's second law of diffusion has already been described in relation to coffee extraction (Chapter 5) and drying (Chapter 6); but it is questionable whether its mathematics for unsteady state diffusion is applicable to the release of CO_2 from a coffee bean, since release or diffusion is occurring under pressure.

Nevertheless, Heiss[1,2] gives an equation for CO_2 release of the form

$$N = A(B - P_{CO_2})[1 - (Ce^{-Dt} + Ee^{-Ft})]$$

where N is the flux of CO_2; P_{CO_2} is the external partial pressure; t is time; and A, B, C, D, E and F are constants. This equation is very similar to that from integration of the diffusion equation (see Chapter 6, section 2.6.3) in its exponential form. The negative effect of P_{CO_2} is of interest, and a graphical plot is shown in Fig. 1. As with the diffusion equation, a plot of log E vs. t would show a linear relationship up to about $E = 0.5$, at a fixed value of P_{CO_2}. For the degassing of coffee, more fully discussed in sections 3.1 and 5.1, we are interested in the evolution when $P_{CO_2} \approx 0$; whereas the evolution in a closed container will require a differential solution of the above equation since P_{CO_2} is also changing with time. For the latter reason, the data of Radtke given do not conform to a linear

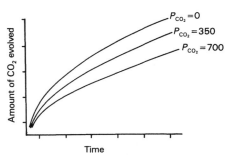

Fig. 1. Typical course of CO_2 evolution from roasted coffee, relating cumulative amount evolved with time at different external partial pressures of CO_2 (mmHg abs.) after Durichan and Heiss.[1]

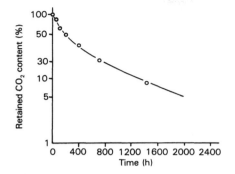

Fig. 2. Relationship of percentage retained CO_2 content (of initial amount) on a log scale, with time of holding in a closed package (h). Calculated from data of Radtke.[4]

relationship of $\log E$ with t, as shown in Fig. 2, for the roasted coffee sample No. 1 from Table 1 (initial amount of CO_2 at time zero taken as 378 ml per 100 g).

When, therefore, roast whole beans are placed within a closed pack, the CO_2 as it is released will have to be contained within the available headspace within the package (which will include the space between the beans or their external porosity); together with the excess volume provided over and above that of the specific volume of the beans themselves). A portion may also diffuse through the film material of the package. There will be a resultant partial pressure of CO_2 inside the package, which will control to some extent the release of CO_2, and a total pressure, the final value of which may or may not be acceptable for the type of package. A figure for the total overpressure in a closed impermeable pack starting at atmospheric pressure may be derived from the internal volume of the pack, say 10% over that of the roast whole beans, at, say, 330 ml for 100 g coffee (of which about 150 ml will be occupied by the beans themselves, but only 75 ml by the solid material of the beans at 0·50 internal porosity). There will be a minimum $366 - 150 = 216$ ml total headspace available for take-up of released CO_2. Every 100 ml introduced will increase the total internal pressure by $100/216 = 0.46$ atm.

The possible solutions to this problem, where the release of up to 500 ml CO_2 are clearly possible from 100 g coffee, are (1) to have a sufficiently rigid container—impractical; (2) to pack in a bag of material sufficiently permeable to CO_2, or fit the bag with a one-way valve; (3) to pack under vacuum—though even a full vacuum may well be inadequate; (4) to pack with a very large headspace—also impracticable commercially; and (5) to allow substantial release of CO_2 by holding the roast beans before packing.

2.2. Stability Factors

There is a further problem in packing, though not so great as with roasted and ground coffee. 'Staling' of coffee is a well known phenomenon, resulting from the action of the oxygen of the air on the components of the roast coffee (and other interactions between components), increased by increased temperature and moisture content. Roast coffee will also gradually release its volatiles to the atmosphere, some at a greater rate than others, so that in time there would also be both a loss and an unbalancing effect.

The stability of roast coffee has been studied and reported by a number of workers, in particular Radtke-Granzer and Piringer[9] and Cros and Vincent.[10] Published data have been collated by Clarke[11] which show that roasted whole beans, although undergoing continuous but gradual deterioration from time zero, can be regarded as little different from fresh after 10–12 days in an air pack; there is a difference in flavour of a brew after about 40 days, with a limit of acceptability after about 70 days. By packing under vacuum with a low oxygen content, Cros and Vincent indicated a time for significant change as 180 days; though it would be fully acceptable for at least 18 months. They also showed an advantage with a closed metallised pouch rather than an air pack. The effect of temperature was found by Heiss and Radtke[6] to be such that each 10°C temperature increase/decrease decreased/increased shelf life, to the same flavour quality, by 50%, also indicating the advantage of refrigerator storage. There is little information on the precise relationship between the in-pack percentage oxygen content and stability for roast whole beans; but this subject is dealt with in more detail for roast and ground coffee (section 3.2).

2.3. Types of Pack

Roast whole beans are much more stable to oxygen than roast and ground coffee, so that simple bags may be perfectly acceptable for storage (and selling) over relatively short periods of time. Large-scale manufacture and distribution, however, require attention to the problems of stability and of evolution of CO_2. The latter is in fact a greater problem with roast whole beans, due to the larger amount of CO_2 that can be present at the time of packing.

Depending upon the shelf life required, the material of the bags has to be considered in regard to (1) water vapour ingress; (2) oxygen or air ingress; (3) CO_2 and volatile component egress; and, furthermore, the bag material may need to be grease-resistant for the more oily dark-roasted coffees. Permeability data in appropriate units in respect of all of these factors are available for the various usable materials, generally plastic or metal foils. Heiss and Radtke[6] have reviewed these requirements in practice, and find that for short service periods gastightness is not essential (the permeability rate of CO_2 out is generally much greater than that of oxygen in), but that watertightness is important. Multi-layer bags consisting of detached layers of glassine, aluminium foil and regenerated cellulose film have been used, but laminates are now more general.

Unless the bag material is permeable to CO_2, fairly long degassing times could be required. Such packs do, however, have considerable headspace volume in relation to the volume occupied by the beans. The reduction of the initial oxygen content is best achieved by adequate flushing of the bags and contents with an inert gas: nitrogen or CO_2. For longer shelf life, the non-return valves of Goglio or Hesser are popular; that is, a plastic valve fitted to the outside of an otherwise impermeable bag which lifts as required at a pre-set pressure to release excess CO_2. Vacuum packing is not often used though there are such packs, also fitted with one-way valves, for roasted coffee that has been degassed over a period of, say, 4 days.[9]

The mathematics of the situation in respect to net partial pressures of CO_2, O_2 and N_2 within such soft packs and its application has been discussed in detail by Durichan and Heiss.[1]

3. PACKING OF ROAST AND GROUND COFFEE

3.1. Carbon Dioxide Evolution

The release of carbon dioxide from roast and ground coffee has also been studied by Radtke[4] with experimental determination at two different grind levels, one coarse ($\sim 1000\,\mu m$ average particle size), and the other fine ($400\,\mu m$). Selected data obtained are shown in Table 2 taken from the graphical presentation.

It is evident that the release of CO_2 to a maximum is quite rapid, especially with the finer grind as compared with roast whole beans; indeed the meaning of the apparently high evolution at time $= 0$ h is not clear; again there is a problem of time zero, which strictly speaking should be immediately after grinding; though for packing studies the zero is immediately after filling into packages. These figures are taken against a rising partial CO_2 pressure, as for the experimental work with roast whole beans, and do not represent the natural release of CO_2 to the free

Table 2
Release of carbon dioxide from roast and ground coffee[4]

Time from start of measurement (h)	Amount of CO_2 released (ml at NTP per 100 g)					
	Coffee 1		Coffee 1 (decaff.)		Coffee 2	
	Coarse	Fine	Coarse	Fine	Coarse	Fine
0	150	150	200	200	150	150
100	300	200	400	225	325	220
300	—	—	—	—	—	215
400	—	210	425	232	370	—
1100	365	—	—	—	380	—

atmosphere. In natural release, the total amount of CO_2 evolved, including that at grinding, should equal that for the roast whole bean from which they are derived, whatever the degree of grind. It is apparent that substantial quantities of CO_2 are released at grinding, especially with the finer grinds; but it is not clear whether this amount is merely consonant with the rate of release subsequent to grinding. Estimates have been given of the proportion released at grinding from the roast whole bean, e.g. Heiss and Radtke[6] consider that 45% is released within 5 min of grinding for a fine ground coffee, and Barbara[7] determined that 30% was lost within 5 min of grinding to an average particle size of 1000 μm and 70% at 500 μm.

The data in Table 2 again cannot easily be expressed mathematically, since the P_{CO_2} is increasing with time. For a constant external P_{CO_2} the evolution is expected to follow an exponential equation of the type already described in section 2.1. Experimental studies on natural evolution will be interpretable by plotting $\log E$ against time. It is probable that initially, up to a value of $E = 0.5$ (or 50%), the relationship between E and t will be linear.

Some other experimental work is published, e.g. reference 12, in which seven samples of a roast and ground coffee (drip grind, at an average particle size of 810 μm) packed 1 h after grinding show an average of a total $550/454 = 121$ ml of CO_2 per 100 g after 40 days: figures somewhat lower than in Table 2. This paper includes data on the composition of the evolved gas, when also 0.002 ml of O_2 per gram of coffee were reported to be emitted and shown to be fully sorbed after 40 days. Saleeb[13] provided adsorption–desorption curves for fresh roast and ground coffee for CO_2, plotting the equilibrium pressure of the CO_2 against CO_2

adsorbed (ml at STP per g) at the temperature of $-78.3°C$, which is of interest relating to the porosity of coffee. A very marked increase in adsorption at about 450 mmHg CO_2 pressure is evident.

3.2. Stability Factors

Stability data from Radtke-Granzer and Piringer,[9] Clinton[14] and others have also been compiled and reviewed by Clarke.[11] The stability of roast and ground coffee during storage is clearly less than that of roast whole beans, so that packing at a low oxygen content is required for a reasonable shelf life. Assessment of flavour quality (that is, after brewing under standard conditions) is necessarily subjective, though use of expert panels and statistical analysis of their findings (ratings on a scale) improves objectivity. According to Heiss and Radtke[6] and Radtke-Granzer and Piringer,[9] three different flavour quality criteria after storage can be set, i.e. 'high' (close to fresh), 'medium' (satisfying) and 'low' (still acceptable without emergence of off or stale flavours). With these criteria, for a closed package at 0·5% initial oxygen content by volume, the corresponding shelf-life times would be 6, 12–17 and 20–25 months respectively; at 1% initial oxygen content, the times are given as 4, 9–17 and 14–20 months respectively. Such figures would apply to a roast and ground coffee stored at around 21°C and having an initial and final moisture content below or around 4% w/w. The same data[9] show the marked decrease in stability for the same roast and ground coffee with oxygen levels up to 5%, whereas a reference sample at zero oxygen content and stored at $-20°C$ showed virtually no change ('high' quality) after 20 months. There is some evidence of a linear correlation of log(oxygen content) v. time at each quality level, including extrapolation to an air peak (21% O_2) with a shelf life of 10–15 days for 'medium' quality. Some differences can be expected for coffees of different grind and roast degree, and blend; and the effect of differing temperature will be as for roast whole beans (section 2.2), which is consistent with an Arrhenius-type equation relating log (rating) to the reciprocal of the absolute temperature for the same storage conditions. The effect of panel ratings in flavour quality from stored and roast coffee at different moisture contents has been studied by Clinton[14] up to 6%. The correct measurement of moisture content presents some problems (see Chapter 4, section 2.5); it is also believed that it is only unbound moisture above, say, a monolayer that can influence stability.

There may also be an effect on stability according to the pre-packing conditions for the roast and ground coffee, that is, its contact with oxygen before packing. Shelf-life tests normally start with time zero at the time of packing; whereas, like gas evolution, deterioration is really starting from immediately after roasting/cooling. Oxygen is believed to be absorbed by roast and ground coffee, which provides a reservoir for oxygen to cause subsequent deterioration. Heiss and Radtke-Granzer[6] remarked on the probability of oxygen pick-up at grinding, and have found evidence for this in that in the vacuum packing of roasted coffee an immediately determined in-pack oxygen value may in fact rise in value after several days as a result of desorption into the vacuum space, followed by its fall again during storage.

Since oxygen is such a prime determinant of shelf stability, there are two main ways of lowering this content (to, say, below 1%) in a commercial pack. The first method is to apply a high vacuum immediately after filling into the package and then closing; the other is to flush the roast and ground coffee and package (before closing) with an inert gas. Consideration is also given to diminishing contact with oxygen before packing.

In consumer use, Clinton[14] pointed out that commercial packages will generally be sold on average three months after factory packing, and most unlikely before a month; though much longer may be experienced with slow-moving stock. Clinton also showed that representative US consumers were not able to detect many changes detected by experts. In consumer use, packages will be opened and reclosed frequently, and the shelf life is essentially that of an air pack. An advantage is to be gained by actually closing the pack, rather than letting the contents have direct access to replacing air. A further advantage is to be gained by keeping the contents in a refrigerator provided that the moisture content of the coffee is not increased.

Some compositional parameters have been related to coffee stability, in particular the so-called aroma index, M/B, which is the ratio of the content of 2-methylfuran to that of butanone, measurable by GC techniques, as used by Kwasny and Werkhoff,[15] Vitzthum and Werkoff[16] and Arackal and Lehmann.[17] This technique is useful for a roasted coffee of known blend and roast degree, to which the M/B ratio will relate in a known manner. On storage, this ratio falls from, say, 3·2 for a given coffee when fresh, down to, say, 2·3, at which point a loss of coffee flavour is noted; M/B values were correlated with a flavour rating scale.[16] The technique cannot, however, be used to assess the condition of a coffee of an unknown type. Other indicators have been examined by Radtke-Granzer and Piringer[9] and by Tressl and Silwar.[18]

The shelf-life factors above are primarily related to flavour quality measurement of the subsequently brewed coffee. Headspace aroma from the dry product is of some consumer significance, and although data are not readily available, decline in this aroma will be slightly faster than reduction in flavour quality. However, a decline in this headspace aroma does not necessarily mean a corresponding deterioration in the flavour quality or headspace aroma of the brew. It appears that the volatile compounds in coffee do not have the same significance for each attribute, i.e. certain compounds are more important for headspace aroma from dry product than for flavour in a brew, and vice versa. Although the use of in-pack percentage O_2 content figures shortly after packing is widespread, and also that of in-pack vacuum measurements by special gauges, measurement of content by weight (of O_2) per unit weight of coffee is regarded as more satisfactory. Such a maximum figure in a commercial pack might be 20 µg per gram of coffee, including that which has been absorbed, corresponding to 1% oxygen content in a high-vacuum pack (immediately after desorption) for satisfactory shelf life. Radtke[5] describes a method of measurement of oxygen content in the headspace in µg O_2 per gram of coffee. Heiss and Radtke[6] prefer to consider the partial pressure of oxygen in the pack as being the key parameter for stability (e.g. $P_{O_2} < 10$ mbar). An often quoted figure from 1947 for the amount of oxygen required to 'stale' roast coffee (i.e. to the point of non-acceptability) is 14 ml per pound of coffee; this figure calculates to 0·031 ml per gram or 40 µg per gram, though such figures do not indicate in themselves the rate of staling. As already suggested, a logarithmic relationship of initial oxygen content with time to a given quality level is likely, so that 20 µg of oxygen may also bring coffee to staleness, but in a much longer time. Nevertheless, there may indeed be some minimum oxygen content required to bring coffee to complete staleness before it is all reacted, but scientific data are lacking. Temperature, again with a probable logarithmic effect, is also an important parameter in stability.

3.3. Types of Pack
The types of pack and packing methods used need to take into account the considerations outlined in the two foregoing sections.

3.3.1. Cans
The oldest type of commercial pack is the vacuum-packed metal can (i.e. tin-plated steel with lacquer covering). Cans are of standard sizes in the

USA and the UK. Sivetz and Desrosier[19] mention that the most popular size of can in the USA for roast and ground coffee is that designed for holding 1 lb of coffee, measuring $5\frac{1}{8}$ in outside diameter and $3\frac{1}{2}$–$3\frac{5}{8}$ in high, with a volume of about 66 in^3 (1·08 litres). Cans have considerable advantages for packing coffee in that they fulfil a number of useful criteria, notably impermeability to water, air and volatile compound egress or ingress; furthermore they can be directly lithographed without need for labels. They have, however, to be sufficiently rigid to withstand the initial high vacuum required, which is of the order of 28·5 inHg (against 30 in barometric pressure), and preferably higher, to give less than 1% oxygen in the can, and therefore the coffee stability required. As a result of CO_2 gas evolution from the roast coffee, this vacuum will be reduced until atmospheric pressure is approximately restored, but no more. It is therefore necessary to have the roast and ground coffee at a CO_2 content that will fulfil this condition, by pre-release or degassing in bulk as necessary.

The type of calculation required is illustrated by the following example. Suppose, with the 1080 ml capacity can already mentioned holding 454 g coffee, the measured in-can percentage oxygen content by volume taken or converted to atmospheric pressure (760 mmHg = 30 inHg = 1013 mbar = 101·3 kPa) is 1·0%, measured shortly after packing. The partial pressure of the oxygen will be 10 mbar. If the headspace in the can initially is essentially air, i.e. a nitrogen to oxygen ratio of 79:21, then the absolute pressure inside the can measured at the same time should be $1·0 \times 30/21 = 1·4$ inHg, or a so-called vacuum of 28·6 inHg (against an atmospheric pressure of 30 inHg). If the headspace in fact contains substantial quantities of other gases such as CO_2, then for the same oxygen content a higher absolute pressure or lower vacuum is possible. The headspace volume may be calculated by subtracting from the total can volume the volume occupied by the roast coffee particles, for which a typical figure would be 1·4 ml per gram of coffee; i.e. per gram of coffee, $1080/454 - 1·4 = 0·98$ ml. Taking the density of air at 20°C and atmospheric pressure as 1·331 g per litre, then the actual weight of oxygen in this headspace is $1·331 \times 10^{-3} \times 1 \times 10^{-2} \times 0·98 = 1·30 \times 10^{-5}$ g or 13 μg per gram of coffee. The total headspace should include, strictly speaking, the internal pores of the coffee, which at a porosity of, say, 0·49 provides an additional volume of $0·49 \times 1·4 = 0·69$ ml per gram of coffee. These pores will contain both CO_2 that is being released, and perhaps also occluded the sorbed oxygen. It is unlikely, however, that this volume of gas will be abstracted during a short evacuation time of the vacuum packing process, and it will therefore constitute initially a gas of different composition from that now in the main headspace. Equilibration will, however, take place inside the can, so that, according to the oxygen level within those pores, the measured oxygen content within the can will rise (offset by any loss of oxygen by reaction with the coffee in that time). That this can be so has been demonstrated,[5] e.g. a rise to 1·1–3·0% from 1% after about 2 days, followed by fall as the oxygen reacts to cause staling of the coffee. The measured in-pack content of oxygen can then be used in conjunction with the total headspace volume to determine the weight of oxygen present in μg oxygen per gram of coffee. Measurement of the in-pack percentage nitrogen content at any stage is of value (provided that no flushing with nitrogen has ever taken place), since this will have been associated with the oxygen in the ratio 79:21 (except that it is believed that a small percentage of nitrogen is associated with the evolved CO_2). A comparison with the direct oxygen content will indicate the adsorption or reaction with the roast coffee.

Knowledge of the available headspace volume in the can is important, since with the initial vacuum applied it determines the allowable residual CO_2 in the roast coffee to be packed. In this instance, 0·98 ml per gram of coffee is available (and perhaps more, if the internal headspace is also considered, i.e. up to 1·50), and the coffee to be packed should have the potential for release of not more than this amount if atmospheric pressure in the can is not to be exceeded. However, this potential cannot be assessed from data of CO_2 evolution to atmosphere, even if available, since evolution will be suppressed somewhat by virtue of the increasing partial pressure of CO_2 within the can. After grinding, a fine grind medium roast arabica coffee may well only contain 1·1 ml per gram, and a coarse grind 1·5 ml per gram, which may be fully released to a low partial pressure of CO_2 as in air, but perhaps only three-quarters of this amount under the conditions inside a can. In these instances, prior degassing for any period beyond about an hour would not be necessary, as is found in practice. Calculations do not, however, replace the need for simple trials in particular cases to determine the degassing time required.

Cans may also be used for holding roast and ground coffee in an inert gas atmosphere. The usual procedure in this case is first to apply a vacuum (which need not be high), and release to a supply of inert gas such as nitrogen. In order, however, to prevent excessive pressures in the can, it will be necessary to degas the coffee to a much lower level than for high-vacuum packing. Alternatively, a relatively low vacuum can be applied to the gas-purged can before finally closing.

3.3.2. 'Hard' Packs of Laminates

As an alternative to the can, so-called 'hard' or 'shape-retentive' packs of flexible laminated materials shaped in bags have been developed and widely used in recent years in Europe. These packs are called 'hard' since on application of a high vacuum after filling and sealing, the material collapses on to the coffee to form a 'brick' hard to the touch. Any substantial evolution of CO_2 within the bag, or ingress of air through leaks, will render the bag 'soft' and is an undesirable condition. 'Softness' does not imply any deterioration of the coffee on that account, but may well be perceived by the consumer as such.

The bag materials consist of laminates usually of about three different layers, of which aluminium foil is usually the central for moisture ingress protection, with a heat-sealable inner layer of plastic material such as polyethylene; there is a recommended use[6] of PETP at 12 μm thickness, aluminium at 12 μm and LDPE at 70 μm. The sealed filled bag is then usually placed within a closely fitting cardboard box for retail sale.

The vacuum employed just prior to sealing is similar to that used for cans, i.e. the objective is to provide an in-pack oxygen content of 1% or less. In such packs, however, the available headspace for CO_2 evolution is much less, so that for packing 1 lb of coffee, a package volume of, say, 930 ml might be used (instead of 1080 ml). Furthermore, it is no longer permissible to allow the internal pressure on storage to rise to atmospheric pressure or beyond. In fact, it is necessary to assess at what internal pressure the packs will become noticeably soft, which will be of the order of $\frac{1}{2}$–$\frac{3}{4}$ atm. Heiss and Radtke[6] consider that the pressure should not rise to more than 735 mbar (i.e. 0·725 atm = 551 mmHg). The development of pressure in such a gastight pack was examined by Radtke[3] for six freshly roasted coffees, in combinations of light to dark roast degree and fine to coarse grind. Graphical presentation of results, only with dark roast/coarse grind, medium roast/coarse grind and dark roast/coarse grind, indicated that the pressure limit was exceeded; the same result was also shown for medium roast/medium grind, dark roast/fine grind and medium grinds from numerical data. Data on actual evolution of CO_2 in ml per gram of coffee were not given. According to the method of Heiss and Buchner,[2] the necessary degassing times can be determined and will range from 1 h to 30 h; for example, medium roast/coarse grind will require 8 h.

These kind of packs are, however, somewhat more susceptible to leakage than vacuum-packed cans, because of the potential for 'pin-holes' in the laminates and any imperfections at the seal.

3.3.3. Soft or 'Pillow' Packs

These packs (also called pouches) are of the same kind that have been described for use with roast whole beans, and the same considerations apply. The packs are described as 'soft' since no vacuum is used, and the reduction of the oxygen content is carried out by use of as efficient inert flushing techniques as possible.

There is a wide variety of such bags of different plastic/metal foil laminates, capacities and shapes, which are described by Heiss and Radtke[6] and by Sivetz and Desrosier.[19] Again, a certain level of permeability by the bag material is acceptable, though gas release is less than with roast whole beans, as has been described. However, protection against oxygen/air/moisture ingress with a low initial oxygen content is especially important for a good shelf life. The seals must be good, with a possible problem from roast coffee dust entrapped by electrostatic forces, and there must be no pin holes in the bag material.

The extent of degassing that may be required will follow from experimental trials; but the methods of assessing the development of each of the partial pressures of carbon dioxide, oxygen and nitrogen of Heiss and Buchner[2] should be examined.

4. PACKING OF INSTANT COFFEE

The problem of gas evolution does not exist with instant coffee, but that of ensuing stability over a reasonable shelf life (say a year to 18 months) remains. There are, however, only a very limited number of references to published studies of stability in storage, notably those of Harris et al.,[20] Clinton[14] and Dart;[21] though there is much information of a trade journal nature.

Commercial instant coffees differ very widely in character, resulting not only from blend and roast degree and solubles yield taken, but more importantly from the amount and nature of the retained volatile compounds following the numerous types of process routes that can be taken, as described in Chapters 5 and 6. Increased sophistication in processing is now general, compared with the early simply spray-dried powders, and this is reflected in the much higher measurable (by GC determination and flavour assessment) volatile compound content. Such aromatics, as with roast and ground coffee, are susceptible to flavour deterioration from oxygen/moisture pick-up. Certain commercial instant coffees have a surface application of coffee oils carrying aromatics particularly responsible for headspace aroma, which again are susceptible to deterioration.

There is also a basic need for the prevention of any moisture ingress into the packed product, since, apart from any other considerations, instant coffee at about 7–8% moisture content will 'cake' into a pasty or solid mass. An inspection of water sorption isotherms (Volume 1, Chapter 2) will show how readily this can occur even from air of quite low relative humidity.

The data of Clinton[14] were based upon two commercial instant coffees in sealed jars, one of granules (spray-dried/agglomerated) with an initial and constant moisture content of 4·3% and an initial in-jar oxygen content of 4%; and the other freeze-dried with an initial and constant moisture content of 2% and an initial in-jar oxygen content of 2%. The latter in particular showed quite a marked drop in acceptability by expert-panel ratings in the first few months but a slower rate of deterioration at a still acceptable level thereafter, which was slightly increased by raising the moisture content to 5·1% and 3·2% respectively; however, the product was still acceptable at 18 months. Although it is difficult to generalise for all instant coffees, it appears, therefore, that with such products the in-jar oxygen levels should be kept down to values comparable to those used for packing roast and ground coffee. With many instant coffees, it is well known that simple spray-dried coffees (provided the moisture content is maintained at less than 4–5%) will retain their original quality for at least two years under ambient conditions, even though the headspace in the jar is at 21% oxygen content (i.e. an air pack).

The evaluation data on shelf life at 37°C by Harris et al.[20] are unsatisfactory for a number of reasons. However, the changes of headspace composition with time are of interest, in particular the increase in CO_2 percentage due to slight internal compositional changes at this temperature.

Instant coffee was once entirely packed in tins for the retail market, but around 1960 the use of glass jars of various shapes became popular. Moisture-proofness is ensured by a waxed paper or metal foil diaphragm well sealed to the rim of the jar; screw caps of plastic or metal finally close the jar. The final assembly is intended also to ensure no ingress of oxygen to the jar headspace when the jar has been gas-packed to a low oxygen content.

5. PACKING EQUIPMENT

5.1. Degassing Plant

When degassing for a particular pack of roast coffee is thought necessary, the coffee after roasting or after roasting and grinding is generally held in containers, e.g. tote bins, etc., for the necessary time; moderate heat may even be applied so that the contents are kept at, say, 30°C to hasten the process. Arrangements should of course be made for venting away the evolved gases.

One degassing system employed commercially involves the use of a vacuum which is applied to the coffee held in closed containers with valving arrangements, followed by inert gas introduction and a second evacuation (by vacuum). Finally, the container is flushed again with inert gas to a pressure slightly in excess of atmospheric, and the contents are discharged to the filling machines. This system (method and equipment) was first patented in the UK with an application date of 1962. Table 3 gives a listing of some of the relevant patents. These patents claim that the degassing time is shorter than otherwise possible at atmospheric pressure, that the initial evacuation removes any occluded oxygen from within the ground coffee particles, and that the pressure in the container (around 500 mmHg absolute) after the inert gas has been introduced should be that of the 'vapour pressure of the coffee' (i.e. that of the more volatile compounds, so that these are not lost during the degassing). The significance of vapour pressure of pure volatile compounds is not clear in this context as partition coefficients are more meaningful, nor are there any GC data offered as to actual losses of these compounds. According to the analysis of degassing by Heiss and co-workers[1,2] already discussed, there will, however, be a reduction of degassing time provided that the

Table 3
Selection of vacuum degassing patents

Patent No.	Date of publication	Assignee
BP 1,024,314	1966	Kenco Coffee Co., London
USP 3,333,363	1967	
BP 1,200,635	1967	
USP 3,506,446	1970	
USP 3,613,549	1971	
Can.P 853,634	1970	

partial pressure of the CO_2 in the ambient environment of the ground coffee is kept as low as possible.

From the fundamentals of degassing, there would also be an advantage in agitating the contents of a degassing bin to reduce local concentrations of CO_2.

5.2. Roast and Ground Coffee
Both Sivetz and Desrosier[19] and Heiss and Radtke[6] provide descriptions of packing equipment for this purpose. The types of packing equipment for soft packs vary in their speed of operation, with speeds of 70–120 packs per minute now being looked for, and usually incorporate all form–fill–seal and flushing operations. In Europe, Hesser of Stuttgart, Germany, and SIG of Switzerland, and in the USA, Bartelt of Illinois, are especially well known manufacturers.

The 'hard' pack of flexible laminates using high-vacuum packing was first introduced about 1960 in the 'Ceka container' system of Christenssons of Stockholm, Sweden, though there are now others.

Vacuum packing of coffee in cans following their initial filling was originally carried out in machines with rotary movement to 'clinching' (lids), vacuumising, and final sealing stations, before passing to conveyors for subsequent operations.

5.3. Instant Coffee
Relatively little information[19] is generally available on packing machines for instant coffee, which have now reached considerable complexity and are required to be run at very high speeds, up to 360 jars per minute for the smaller sizes. Volumetric fill at stations on turntables, but with the application of an adjustable vacuum to give the required weight according to bulk density, is usual, for example in machines supplied by the Pneumatic Scale Corporation of the USA or the Albro Fillers Company of the UK.

Additional complexity arises from the need to gas-pack the jars at a definite in-jar oxygen level, and machines may also incorporate systems of spraying coffee oil on to the instant coffee immediately before filling. More expensive machines with direct weighing have also been manufactured by the Pneumatic Scale Corporation.

5.4. Weight Control
An important aspect of filling both roast and instant coffee is that of weight control to meet strict legislative requirements in the USA, the

EEC and other countries. At the same time, both products being relatively expensive commodities, the manufacturer will need to ensure minimum 'give-away' over the legislative requirements of in-jar weight, and so will require accurate control systems.

REFERENCES

1. Durichan, K. and Heiss, R., *Verpackung-Rundschau*, 1970, **21**, tech. wiss. Beilage, 35–41.
2. Heiss, R. and Buchner, N., *Verpackung-Rundschau*, 1963, 67–74.
3. Radtke, R., *Proc. 6th Coll. ASIC*, 1973, 188–98.
4. Radtke, R., *Proc. 7th Coll. ASIC*, 1975, 323–34.
5. Radtke, R., *Proc. 9th Coll. ASIC*, 1980, 81–98.
6. Heiss, R. and Radtke, R., *Proc. 8th Coll. ASIC*, 1977, 163–74.
7. Barbara, C. E., *Proc. 3rd Coll. ASIC*, 1967, 436–41.
8. Treybal, R. E., *Mass Transfer Operations*, McGraw-Hill, New York, 1981.
9. Radtke-Granzer, R. and Piringer, O-G., *Deutsche Lebensm. Rundschau*, 1981, **77** (6), 203–10.
10. Cros, E. and Vincent, J. C., *Proc. 9th Coll. ASIC*, 1980, 345–52.
11. Clarke, R. J., in *Handbook of Food and Beverage Stability*, Ed. G. Charalambous, Academic Press, New York, 1986.
12. Cartwright, L. C. and Snell, C. T., *Spice Mill*, 1974, **70** (3), 16 and 24.
13. Saleeb, F. Z., *Proc. 7th Coll. ASIC*, 1975, 335–9.
14. Clinton, W. P., *Proc. 9th Coll. ASIC*, 1980, 273–86.
15. Kwasny, H. and Werkhoff, P., *Chem. Mikrobiol. Technol. Lebensm.*, 1979, **6**, 31–2.
16. Vitzthum, O. G. and Werkhoff, P., *Chem. Mikrobiol. Technol. Lebensm.*, 1979, **6**, 25–30.
17. Arackal, T. and Lehmann, G., *Chem. Mikrobiol. Technol. Lebensm.*, 1979, **6**, 43–7.
18. Tressl, R. and Silwar, R., *Chem. Mikrobiol. Technol. Lebensm.*, 1979, **6**, 52–7.
19. Sivetz, M. and Desrosier, N. W., *Coffee Technology*, AVI, Westport, Conn., 1979.
20. Harris, N. E., Bishop, S. J. and Mabrouk, A. F., *J. Fd Sci.*, 1974, **30**, 192–5.
21. Dart, S. K., PhD thesis, University of Reading, 1985.

Chapter 8

Home and Catering Brewing of Coffee

G. PICTET

Linor—Food Development Centre, Nestec Ltd, Orbe, Switzerland

1. INTRODUCTION

The proportion of food products and beverages offered to the consumer that are prepared on an industrial scale is increasing. However, in certain cases a significant part of the preparation involving various operations is still carried out domestically, or on a small scale. The precise conditions in these operations, as well as the equipment used, can be very important, and a certain skill or know-how may also be valuable.

During industrial production of a food or beverage product, two main objectives are sought, namely the production of a high-quality product at the most economic price. These objectives are shared by the consumer who will often have less elaborate equipment available and may sometimes have to cope with problems unknown to the industrial processor. Domestically, criteria of quality (taste, aroma, body, appearance, colour, etc.) may show up very differently from those established on an industrial scale. That is why a sound knowledge of the precise behaviour of the raw materials used in domestic preparations is very useful for the development of a comparable industrial process.

Among the methods used for the domestic preparation of foods, solid–liquid extraction is one of the more common, especially for beverages and infusions (medicines, tea, maté, coffee, etc.). The recent appearance of several household types of equipment of novel or improved construction has stimulated a renewal of interest in the brewing of roast and ground coffee. In this particular case the existence of numerous reference books, both fundamental and technical, has encouraged several studies in recent years. In the following sections we shall review the results of the more interesting of these studies and then attempt a description of our own work in this field, as presented at the 8th ASIC Colloquium.[1]

2. BIBLIOGRAPHIC REVIEW

2.1. Solid–Liquid Extraction
2.1.1. Definition
Certain liquid foodstuffs come from a solid vegetable raw material (fruits, vegetables, stimulant foods, etc.), which is submitted to a preliminary grinding and is then contacted with a suitable liquid. The practical conditions of this contact (temperature, pressure, relative proportions, length of time) are established in such a way that the soluble constituents of the treated material pass into the liquid phase in order to form the required brew. The physical phenomenon causing the passage of the soluble constituents into solution is solid–liquid extraction, a process frequently applied in food technology. Several examples of the utilisation of this process have been the subject of recent publications, which refer sometimes to a specific raw material, for example, sugar beet cosettes,[2] leaf tea,[3] roast coffee beans,[4] green coffee beans,[5] and various fruits and vegetables.[6]

Other publications mainly concern the available techniques for solid–liquid extraction, such as those published by Leniger and Beverloo,[7] Spanninks[8] and Schwartzberg[9,10] (see also Chapter 5). These latter papers are mainly concerned with application on the industrial scale of results obtained in the laboratory.

2.1.2. Description
In his PhD thesis, Clo,[11] in providing an excellent résumé of the papers enumerated above, gives a very complete description of solid–liquid extraction, which can be summarised in the following way. The overall phenomenon of solid–liquid extraction is composed of a succession of discontinuous stages, as follows.

Initial state. It is assumed that the solid component to be extracted is distributed in a more or less uniform way in the solid phase. At the beginning of the process, this solid phase is placed in contact with the solvent, which permeates the matrix and dissolves the soluble component.

If this first stage takes place very rapidly, a precise state can be established at time zero of the extraction:

— the solvent in the interior of the solid has a very high solids concentration;
— the solvent at the exterior of the solid has a concentration near or equal to the initial concentration.

Transfer period. A transfer of mass is initiated almost instantaneously in the very concentrated solution contained in the solids, towards the very dilute solution existing at the surface of the latter. When the basic material is a fresh product (fruit or vegetables), the first part of this transfer is influenced by any intervening modifications made to the product. Even in this particular case, the essential part of the transport is explained by molecular diffusion of the solute (material extracted) in the solvent.

Final state. As soon as the concentrations on the interior and exterior of the solid are identical, a state of equilibrium is reached. The initial solvent enriched with solute is called the 'extract'; the insoluble part, depleted of the extracted solute but augmented with imbibed solvent, is described as 'raffinate'.

Certain papers[9,12,13] have illustrated the amounts of the three parts (insoluble fraction, soluble fraction and solvent) in the form of a graphical presentation called a ternary diagram. At each instant the composition of a mixture can be designated by a point situated in the interior of the triangle, each of the apices of which represents one of the three parts in the pure state. In order to obtain the final state expected, it is possible to interpose the transfer stage on the initial state and the conditions used in practice.[14,15]

2.2. Brewing of Roast Coffee
2.2.1. Manual Techniques

The manual techniques used for the preparation of a cup of coffee have been reviewed by various authors,[16-20] and the most frequently used among them can be described in the following manner.

Filtration (percolation). The ground coffee, contained in a filter of paper, cloth, porcelain or other material, is extracted by boiling water. Preparations of this type depend more than others on a particular knack or know-how for each consumer. Heiss and Radtke[21] recommend a preliminary wetting of the bed of coffee, before a normal extraction or lixiviation of the ground material. In the USA this operation is carried out in 'drip-pots', or 'filter cones' in the UK.

Infusion (properly so called). For this method, utilised more commonly for brewing tea, the roast and ground coffee is contacted with boiling water in a suitable receptacle, left to stand over a period of 5 to 10 min, then decanted through a screen or through a filter in order to separate the 'raffinate' from the required extract. In the USA this operation is carried out in open pots, or jugs in the UK.

Boiling (decoction). This method is similar to the preceding, except that the coffee in suspension is heated during the whole of the infusion, a time which is also shortened to avoid the formation of undesirable sensory characteristics. In the UK this method would be called the saucepan method.

Turkish coffee. This is a particular variation of the decoction technique, which necessitates the use of specific equipment for grinding and decoction of the coffee. The latter has to be particularly finely ground, and its suspension in water has to be previously mixed with powdered sugar and then heated strongly. There exist a large number of recipes for the preparation of Turkish coffee.[22]

2.2.2. Household Equipment

Since the beginning of the century, very many types of domestic equipment (or coffee brewers) have been developed to facilitate the task of the housewife and that of the individual consumer. These various types of equipment, which function in a similar manner to the manual methods described in the preceding section, have also been the subject of more or less detailed descriptions.[1,17,18,23-26] In spite of the large number of different types of equipment commercially available, it is possible to divide them into three distinct categories according to their techniques of use. (The various types of equipment are, however, often named in common usage in the UK, USA and other English-speaking countries somewhat differently, as will be mentioned. Some types of equipment are often called after the brand name of the manufacturer.)

Filtration under gravity. The water required for the extraction, brought to boiling by means of a heating element, is forced by a slight excess pressure over the ground coffee placed in a filter (of glass, porcelain, metal or other material). The extraction of this coffee takes place in a continuous manner, until the total volume of the liquid taken for the operation has been used. There exist several variants of filtration under gravity, notably filtration under pressure, and filtration with recirculation, as used in so-called 'percolators' in the UK and USA, either electric or stove-top; these contrast with so-called automatic filter machines or brewers, which will be examined in a more detailed manner in the following pages.

Continuous water circulation (or contact). This is an application of the boiling technique, which operates by means of an apparatus or brewer in which the ground coffee is kept constantly in suspension in the extract by boiling water entering in from a second chamber. At the end of the process, the cessation of water vapour production and the cooling that ensues cause a relative vacuum which facilitates the separation by filtration of the exhausted coffee from the extract ready for drinking. This equipment is often referred to as a vacuum-maker in the USA, and by the brand name 'Cona' in the UK.

Vaporisation under pressure. Earlier we mentioned that filtration can be practised under pressure with suitably modified equipment. This type of equipment is particularly suited to process coffees, highly roasted and finely ground, and to the preparation of a very concentrated drink ('espresso' and 'ristretto', according to the designations used by Italian consumers). Another category of equipment causes the extraction of ground coffee by a mixture of boiling water and its vapour, which takes place against gravity; this is also equipment of Italian origin, and its function will be examined more deeply in our personal studies (section 3.1).

2.2.3. Catering-type Equipment

Equipment designed for large numbers of consumers at a time (i.e. catering) has not been the subject of many published papers.[26,27] It is mainly in the work of Sivetz and co-workers[17,18] that one can find usable information reporting this type of equipment, the functioning of which is, however, very similar to that of equipment described in the preceding sections.

Urns. This is an apparatus working on the principle of filtration by gravity for the preparation of a large number of cups, 60 or more in one percolation. The brew obtained is kept at drinking temperature by the presence of a double-walled shell filled with water and suitably heated; a take-off fixed to the base of the vessel permits nearly continuous supply.

Equipment for airlines. This type of equipment also uses the principle of filtration, but it operates under a slight overpressure and assures the preparation of a litre and a half of brew every 3 min. The technical advantage of this extraction device is that it may be controlled in a regular and scrupulous manner for the production of a product of above average quality according to Sivetz.

Vending machines. These entirely automatic machines have been developed over 30 years and are now to be found in company restaurants, canteens and offices. These machines are designed to provide the customer, no matter what the time of the day, with a cup of hot beverage, with added milk and sugar as required. An important group of these machines use instant coffee, which excludes them from our centre of interest.

Others use roast and ground coffee, introduced by small individual portions in a paper filter, each of these portions being detached for the immediate preparation of a cup of the beverage. Some machines of very elaborate construction have recently appeared on the market, in which the production of a cup of coffee operates in its entirety from the grinding of the roasted coffee beans to the provision of the finished infusion and the elimination of the coffee grounds. As with the preceding types, these machines work by filtration under gravity, sometimes under a reduced pressure.

Espresso machines. These Italian machines, working by application of the technique of vaporisation under pressure, have obtained very great success, initially in the greater part of European markets and then in the majority of markets external to Europe. This success has favoured the development of espresso machines with multiple percolation heads, designed to be placed mainly in coffee bars or in restaurants.

2.2.4. Raw Materials

For the preparation of a brew of roasted coffee, only two raw materials need to be taken into consideration; the ground roasted coffee and the water for extraction. Other materials can, however, take part in this actual operation, and we will review them briefly in the section concerned with additions.

Roasted coffee. The botanical, physical and chemical characteristics of this raw material, as well as the different processes that lead to its preparation in the producing countries, have been described by Coste[28] in an extremely complete manner; see also this volume, Chapters 1 and 2. Techniques permitting the determination of the species and varieties constituting a mixture have also been discussed repeatedly.[29-31] The possibility of submitting green coffee beans to a pre-treatment such as dewaxing[32] (see also Chapter 2) or to irradiation with γ-rays,[33] with the object of improving the properties of the roasted coffee and those of the corresponding cup of coffee, have also been considered.

The development in the course of roasting of the gustatory or taste and aromatic characteristics of coffee is of primary importance for the sensory qualities of the brew.[34] The conditions during this essential unit operation (roaster type, temperature, length of time, pressure, water injection, etc.) ought therefore to be controlled in a very strict manner, as well as the external characteristics of the roasted product[35-38] (see also this volume, Chapter 4). The recent introduction of roasting processes at increased speeds has modified much of the knowledge acquired on the subject of the mechanisms occurring in the course of the operation and the characteristics developing in the raw material.[39,40] Certain claims made by the companies manufacturing this equipment have been questioned by the results of recent workers.[41,42]

Extracting water. Logically, the quality of the water used in the preparation of a brew, and particularly its hardness, is of great importance for the organoleptic (and visual) properties of the latter. The influence of the composition of mineral salts and of organic constituents of this liquid has therefore been the subject of several important scientific studies, mostly of American origin.[43-46] The presence of impurities of all kinds in the water has been examined for the visual effects that their presence can exert on the quality of the resultant brew or beverage.[47] The possibility of modifying the natural composition of mineral salts in the water, by addition or elimination of specific ions, has similarly been considered.[48-50]

2.2.5. Technical Parameters

The physical, chemical and sensory (or organoleptic) properties of a brew of coffee obviously depend upon certain technical conditions in its preparation. The possible interactions of all the conditions (procedure and equipment for extraction, relative proportions of water/coffee, grind degree of the coffee, temperature, pressure, length of time, etc.) have been examined in depth repeatedly.[1,4,51-54] The technical benefits of domestic equipment, which are often most important for the organoleptic quality of the coffee brew,[46] will be studied in more detail in the third section of this chapter.

Clearly the degree of grind of the roasted coffee is primarily responsible for the quantity of dry matter passing into solution in the course of extraction.[55] The effect of grinding on the physical properties of the raw material, e.g. specific surface area, has been the object of a very complete study by the team at ENSBANA at Dijon University,[11,56-58] who have elaborated a very interesting mathematical model to explain the phenomenon of diffusion, the mechanism responsible for the passage into solution of the solid constituents of roasted coffee.

The relative proportions by weight between the ground coffee and the extracting liquid (water) can be considered in two ways. At the time of preparation of a brew using a household brewer, the amount of these two materials is decided above all by the know-how particular to the individual consumer and depends therefore on his own preferences and local customs.[59-61] These personal preferences have been collated through the good offices of the Coffee Brewing Institute (New York City) in the form of a graphical presentation, frequently reproduced in manuals reporting on coffee,[62] which can be seen in Fig. 1. The legal aspect of proportions by weight, and of minimum and maximum concentrations that their variations can cause, is the second factor to consider, and has been discussed by among others Lindner.[63]

The temperature exerted upon the mixture—ground coffee and water—is directly dependent upon the type of equipment used for the procedure, although its influence is not only determined as a direct function of the latter. Some trials, nevertheless, have been carried out in varying this parameter between wider limits than those produced by commercial equipment, that is, from 65°C to 120°C. In a logical manner, it has been demonstrated that the intensity of sensory characteristics, but not their quality, is a direct function of the applied temperature.[46,64] Other research concerns exclusively the effect of the length of time of the operation and is not relevant to the present discussion.

2.2.6. Additions

In most of the papers reviewed in this chapter, the coffee brew prepared by extraction of the ground coffee with water has been considered for its intrinsic nature. It is evident that, following the customs of many consumers, further components are frequently added, which will modify strongly the organoleptic and visual properties of the brew. These additions can be made at the first stage (e.g. use of roasted chicory, roasted cereals, malted cereals, other substitutes, etc.) or at the stage of the final beverage (e.g. use of milk, sugar, other sweeteners, alcoholic drinks, etc.). It is mainly in the addition of milk[65-67] and that of sugar (sucrose)[67,68] that important research has been carried out. The phenomenon of the coagulation of milk proteins or other milk products in contact with the beverage has been particularly considered.

2.2.7. Maintenance of Temperature

Maintenance of the elevated temperature of the coffee brew after its preparation has occupied several groups of research workers. This maintenance evidently seems to prejudice above all the aromatic quality

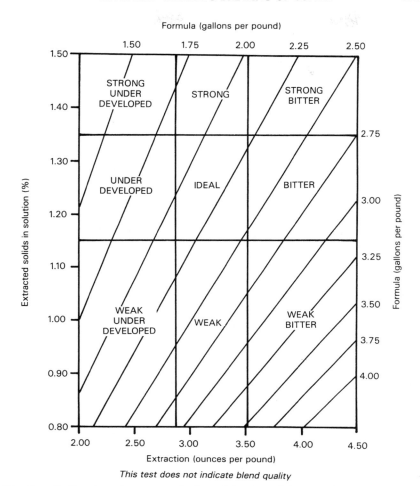

Fig. 1. Coffee brewing control chart (developed by the Coffee Brewing Institute[62]). NB 1 US gall = 3·785 litres = 8·345 lb water; 1 lb = 453·6 g, 1 oz = 28·35 g. Therefore extraction of 3·50 oz per lb = 21·9%. Also, formula of 2·5 US gal per lb = 20·9 water/coffee ratio = 47·8 g coffee per litre of water.

of the brew, since it simultaneously promotes vaporisation of aroma compounds and their chemical degradation.[17,18,69,70] Among the non-volatile components, it is the chlorogenic acids which are mainly involved.[19] But the acids in their entirety are also considerably modified, since a perceptible increase of pH has been observed.[71] The lowering of quality quoted by several of the cited authors above has been doubted by Pangbourn,[46] who for her part has confirmed only slight alterations after 3 h of heating. The effect of a reheating upon various characteristics of a coffee brew has similarly been the subject of a complete study.[72]

2.3. Properties of the Coffee Brew

2.3.1. Sensory Characteristics

The dominant sensory (or organoleptic) characteristics of a coffee brew are as follows: bitterness, acidity and aroma taken in a general sense; as secondary characteristics, it is worth mentioning astringency, 'caramel' and 'phenolic' notes.[73-75] According to Van Roekel,[76] the following taste or gustatory impressions should also be considered: 'paper' notes, harshness, rancidity, saltiness and the 'cereal' and 'sweet almond' notes. Whereas the attributes of acidity and astringency can be attributed to the presence of cyclic and non-cyclic acids,[77] the responsibility for bitterness is not actually imputable to any precise group of components, although possible hypotheses have been proposed for this in recent years.[78-80] The methylxanthines, of which caffeine is by far the most important, do not seem to be as significant as might be expected, because the complexes they form with phenolic acids are organoleptically less active than the free compound.[81]

The phenolic acids, nearly exclusively depsides of quinic acid, should be more correlated with the intensity of astringency rather than that of bitterness.[82,83] Some very recent work in this field seems to provide proof of this correlation, particularly the involvement of the dicaffeoylquinic acids.[84] The 'body' of a coffee brew depends for its part on a combination of organoleptic characteristics, each exerting a synergistic effect on the other.[85,86]

An exact description of the true sensory attributes for a coffee brew is always very difficult, especially when the basis of appreciation is provided by a population of consumers which are insufficiently representative.[87] The establishment of unquestionable correlations between certain organoleptic characteristics on the one hand and certain chemical constituents on the other, for a given infusion, is even more difficult to realise, as it is a question of both aroma and savoury taste properties.[88-90]

2.3.2. Extraction Yield

The terms 'yield of extraction' and 'exhaustion' express the percentage of solid matter of the initial roasted coffee which passes into solution during the preparation of the brew. This extraction yield normally varies between the extreme values of 15 and 35% according to the species/variety of the coffee, the level of roasting, the degree of grind and the technique applied for extraction[91] (see also this volume, Chapter 5). The corresponding extract shows a concentration of soluble solids (taken as dry matter) which oscillates between 1 and 3%; it is convenient also to recall for this purpose that nearly 10% of the dissolved substances may remain in the portion of extract held by the exhausted grounds.

The wide variations mentioned above are in part ascribable to the very large numbers of analytical techniques used for the determination of dry matter content of extract. In spite of the efforts deployed by Navellier et al.,[92] in order to develop a reliable and reproducible reference method, other methods less exact, which are easier and more rapid, have been applied repeatedly in many fundamental studies of unquestionable interest. Most of the researchers involved in the problem of degree of extraction, using well defined equipment or procedures, have published values which are found overall between 9 and 28% as a function of the degree of grind used.[93-96] The highest values in the available references are those mentioned by Ara and Thaler,[97] who report results from a practical extraction under the usual conditions (temperature around 100°C) that reach 40%. See also Chapters 4 and 5 in this volume.

2.3.3. Individual Constituents

Although the total amounts of the soluble components obtained in the course of treatment of the mixture from the infused raw material are undeniably of interest, the effects of the above treatments on each of the individual components are even richer in information. An approximate chemical composition of a coffee brew has been presented by many authors of monographs on the chemistry of coffee, with margins of error and inaccuracies to which such an estimate can give rise.[19,91,94,98,99] The average values suggested by Vitzthum,[91] which unfortunately do not include a mention of the extent of deviation ascribable to technological parameters, have been represented in Fig. 2.

The aromatic principles of a coffee brew are similar, at least in a qualitative aspect, to those contained in the initial roasted coffee product. The quantitative differences recorded between the raw material and the corresponding brew are attributable mainly to the conditions used in the preparation. It is difficult to discern whether these differences are due to a preferential solubilisation rather than to a selective vaporisation.[100,101] According to Streuli,[102] the olfactory impression provided by a roasted coffee should arise above all from its lipid components and the aromatic principles associated with them, which can explain an incomplete transfer of these principles from the roast coffee to the brew, as repeatedly noticed.

From Vitzthum,[91,103] the proportion of caffeine in the original coffee which passes into solution depends markedly on the quantity of coffee used in the brew. The greater the weight of raw material in proportion to the extracting liquid, the less the content of caffeine, expressed as a function of the total dry matter. The transfer of methylxanthines of coffee to the final brew has been the subject of several studies, among which should be mentioned those of Merritt and Proctor,[104] Thomczik[105] and Ndjouenkeu *et al.*[106]

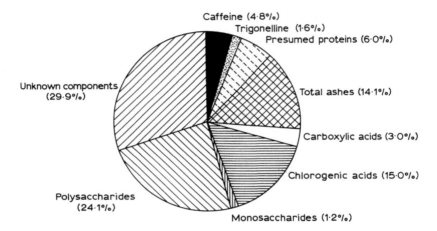

Fig. 2. The chemical composition of a typical coffee brew[91] (percentage dry basis content of named constituents).

Acidity constitutes one of the few fundamental attributes of coffee brews that can be correlated with certainty with analytical quantities. However, the value of the pH of a brew is in practice not affected by the extraction yield in the preparation, as a result of the pronounced buffering effect of the other solubilised constituents. The total acidity, measured as the quantity of base needed to neutralise a given quantity of extract, shows a tendency to increase with an increase in the length of the brewing time[107] and also with an increase of temperature during brewing.[108] The chlorogenic acids, which in some cases represent a minor constituent of a brew, by reason of its degradation in the course of roasting, are composed generally of substances which are immediately soluble.[109-112] The Polish analysts were able upon this evidence to develop a method of estimation based on chlorogenic acids, designed to estimate the strength of a brew and hence the extraction yield for the previous brewing operation.[113]

Total ash content by incineration, and particularly potassium content, have often been proposed as reference constituents for estimating the extraction yield by reason of their good solubility and the reliability of methods used in their measurement.[114-116] Certain of the other metals, often present in trace amounts, have similarly been examined in respect of their solubility;[117-119] however, see also Volume 1, Chapter 2.

The composition, during the course of extraction, of other constituents of coffee with less well defined chemical structure or present in small amounts has also been the subject of fundamental work. Among these constituents it is worth mentioning first trigonelline and its product of degradation, nicotinic acid.[120] The percentages by weight of these two substances in brews depends nearly exclusively on the level of roasting previously applied to the raw material so that their determination in the brew should permit retrospectively the value of this parameter (degree of roast).[121] Nitrogenous substances,[122] pigments[123-125] and constituents extractable with ether[126] have also been examined. As far as free radicals present in coffee brew are concerned, their content, estimated in practice by magnetic resonance methods, shows a tendency to increase in parallel with the temperature used in extraction.[127]

2.3.4. Physicochemical Characteristics

The colour of a coffee brew is often considered as a supplementary attribute of quality; also, its determination by spectrophotometry has occupied several groups of research workers.[128-131] Colour can also be similarly estimated by visual methods by judges participating in sensory evaluation tests;[46] but the use of colorimetric techniques using coloured reference solutions certainly presents appreciable advantages.[132]

Besides oven-drying and rapid-drying techniques, measurement of electrical conductivity of an extract and of refractive index can provide valuable information on the content of conducting substances and total soluble solids;[133,134] see also Volume 1, Chapter 2. Determination of the enthalpy of extracts of robusta and arabica coffee has been carried out by means of an adiabatic calorimeter by varying the temperature of measurement between $-70°C$ and $+80°C$[135] (see also this volume, Chapter 5).

3. PERSONAL RESEARCH

3.1. Introduction

The work that we have carried out in our laboratories over several years had two distinct objectives:

  (a) To define, for certain commercial types of equipment each representative of a particular technique, the influence of the more important operational parameters during the preparation of a coffee brew in the home upon the physicochemical and sensory properties of the resulting brew. The most suitable equipment was to be used in a normal manner for the organoleptic evaluation of the roasted products used in our experiments.
  (b) To determine, with a panel consisting of twelve expert tasters, modifications of sensory characteristics of a household brew, such as are caused by variations of technical parameters already discussed and by the change of one commercial apparatus to another. To specify thereafter, by applying a multi-dimensional mathematical analysis recently developed during the course of our work, those characteristics that can exert a predominant impression upon the judgement of individual experts.

3.2. Experimental Data

3.2.1. Brewing Techniques Compared

From among the numerous commercial types of equipment available, we chose five which appeared to us the most representative of the techniques described in the specialised literature (see section 2.2). Other types working on the same basis have similarly been tested but have provided less reproducible and reliable results and have not been included in this report.

Continuous water circulation (or contact). In the example of the apparatus shown in Fig. 3, the ground coffee is placed in a balloon-shaped vessel positioned above a spherical-shaped receiver in which the water to be used for the extraction is brought to boiling. In the first phase, this boiling water, pushed by the vapour produced in the lower vessel, contacts continuously the ground coffee in the upper vessel. The duration of this phase, which is independent of the equipment used, should be fixed by the personal preference of the consumer.

After this predetermined time, the heating is interrupted causing the start of a second phase of the operation, which consists of a separation of the exhausted coffee grounds from that of the required coffee brew

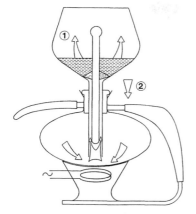
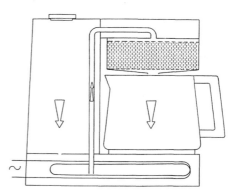

Fig. 3. Equipment for continuous water circulation (vacuum-maker in USA, 'Cona' in the UK).

Fig. 4. Equipment for continuous filtration by gravity (filter brewer (automatic) in USA and UK).

(achieved by means of a filtering plug fixed at the base of the upper vessel). Once the filtration stage has been completed, the upper balloon-shaped vessel and the grounds it contains can be separated from the rest of the equipment and the prepared brew consumed.

Filtration by gravity. In the equipment of Fig. 4, the ground coffee is placed in a suitable filter, and the required water for the extraction in the back part of the equipment is brought to boiling by means of electrical heating incorporated in the base of the latter. Boiling causes a displacement of water to above the filter, then continuous lixiviation (percolation) through the ground coffee, the resultant brew being collected in a receiver positioned directly under the filter. Boiling of the water also assures the maintenance of temperature of brewing during the preparation, which in extreme cases can last some 15 min.

Filtration by gravity under pressure. The principle of the equipment shown in Fig. 5 is very close to that of the equipment previously described, but the modifications made in the heating device, and above all to the fixed system of the filter, permit the process of extraction to take place under a certain excess pressure, caused by the boiling of the water. The extraction process is also strongly accelerated, shortening the time of contact between the ground coffee and the extracting water and thus also the total duration of the operation, which can be reduced under the best conditions to about 100 s.

Filtration with recirculation. In the machine shown in Fig. 6, the

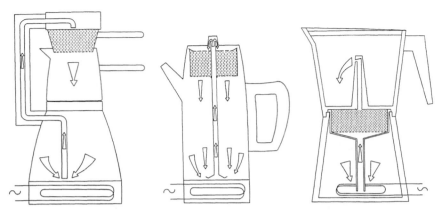

Fig. 5. Equipment for filtration by gravity under pressure.

Fig. 6. Equipment for filtration with recirculation (electric percolator in USA and UK).

Fig. 7. Equipment for vaporisation under pressure (household espresso-type in UK).

extracting liquid (water) percolates in a continuous manner through the ground coffee held in a filter positioned in the upper part of a suitable receiver. Circulation of the liquid, richer and richer in soluble solids, is assured by the thermosyphonic system, fixed directly under the filter. As in the case of the equipment with continuous water circulation first described, the duration of the process does not depend upon the equipment but on the user him- or herself taking into account his or her personal taste. The process is stopped by shutting off the electrical system to interrupt the process of circulation, and the brew thus prepared is ready for consumption.

Vaporisation under pressure. The extracting liquid (water) is brought to a temperature slightly higher than its boiling point and passes through the bed of coffee placed in a filter at the centre of the apparatus after being detached from the upper part (see Fig. 7). The energy necessary for the extraction is provided in this case by an electric heating element, but gas heating is also available on the market. The slight overpressure, depending on the particular construction of the equipment, allows the reduction of the duration of the brewing operation to about 100 s when the maximum volume of liquid is used.

3.2.2. Physicochemical Data

For each of the types of equipment studied, and for the apparatus that we have discarded during our preliminary study, more than 40 brews have been prepared. The raw material used for these preparations was a blend of coffees of good quality, composed mainly of arabica species from Central America and Africa, which had been submitted to a medium roast designed to develop the best taste and aromatic qualities. The technical parameters, which have been varied in the course of this first phase of work apart from the brewing equipment, include the relative proportions of water to coffee and the degree of grind.

For the various liquid brews obtained, for which the exact weight has been measured, the content of total dry matter has been determined by applying the method of oven-drying. This content, based on the weight of ground coffee, has allowed the calculation of the extraction yield in each case. Several of the authors quoted in section 2 have also taken into account, for the estimation of the degree of exhaustion (or extraction), the portion of extract retained by the raw material (or coffee grounds). We have deliberately retained, for the determination of extraction yield, only the fraction available for consumption for each of the liquid brews.

The content of total volatile components of these same brews has also been estimated by measuring the absorption in the ultraviolet region of the condensate arising from the entrainment into vapour of an aliquot of the brew.[136]

The relatively simple checks practised on this first set of samples in the aqueous state have permitted the establishment of certain precise guidelines on the process of extraction, which are shown for each of the sets of apparatus examined. Based on these guidelines, a more limited range of truly representative brews have been prepared in a second phase of the brewing experiments, which have been directly freeze-dried in order to stabilise them. Using these freeze-dried products the content of the main constituents of the original raw material has been determined: residual moisture, by oven drying;[137] caffeine, by the AOAC method;[138] chlorogenic acids, by spectrophotometry;[139] 'protein', by taking the nitrogen content determined by the Kjeldahl method and subtracting the amount corresponding to the caffeine content; total ash, by an incineration method;[140] and saccharides, direct and total, by colorimetry.[141] The analytical techniques used in these determinations of content have been described in a previous publication.[142]

3.2.3. Sensory Evaluation

The representative brews mentioned in the preceding paragraphs were prepared again in order to submit them to a panel of six tasters, who were asked to define their aromatic and taste characteristics. Each of these brews were evaluated, after suitable dilution when it was necessary, by using the technique of paired comparison with a reference brew.

The judges, in evaluating each brew, considered 14 criteria of quality, five relating to aromatic character and nine to taste.

Aromatic: intensity, freshness, balance, heaviness and woodiness.
Taste: intensity ('body'), balance, pungency, bitterness, acidity, astringency, harshness, 'cereal' note and 'earthy' note.

The assembly of these qualitative valuations elucidated by applying a seven-point scale was submitted to a multi-dimensional mathematical analysis. This analysis permitted on the one hand establishment of which criteria were really taken into consideration by the tasters, and on the other determination, among the brewing techniques compared, of the factors that made the brews the most attractive.

3.3. Discussion of Results

3.3.1. Brewing Techniques

In general all the different types of equipment compared have functioned extremely satisfactorily. Any electrical or mechanical incident has not, however, been taken into consideration during our study. Checks on the repeatability of the brewing operation have been carried out, using a medium degree of grind and a relatively high proportion of water to

Table 1
Extraction yield obtained by five different types of brewing equipment

	Brewing equipment	Extraction yield,[a] average and variation (%)
A	Continuous water contact ('Cona' or vacuum maker)	17·9 ± 0·5
B	Filtration by gravity (filter)	22·6 ± 0·2
C	Filtration by gravity under pressure	17·1 ± 0·4
D	Filtration with recirculation (electric percolator)	19·3 ± 0·5
E	Vaporisation under pressure (household espresso)	22·8 ± 0·3

[a] From a medium roast arabica blend with grind of 800 μm average particle size, and brewing ratio of 20 parts water to 1 part roasted coffee (50 g per litre).

coffee. These checks have produced the results in respect of the extraction or exhaustion of the raw material that are given in Table 1.

The limits of variation stated for each brewing apparatus examined are therefore of the same order of magnitude, but a trend in favour of filtration under gravity or vaporisation under pressure is perhaps indicated.

It is possible to determine the quantity of water retained by the exhausted coffee by subtracting from the volume of liquid used at the start that which is present in the collected brew after the process, and neglecting a possible loss by evaporation. These values obtained by subtraction are very close to those obtained by drying the exhausted (or spent) coffee grounds in a vacuum oven. They have been reported in Fig. 8 for those extracts from a medium grind coffee. These values have been compared graphically with a zero retention, represented by the dotted line.

The information that can be drawn from this figure is as follows:

—the quantity of water retained by the ground coffee depends predominantly on the physical characteristics of the latter (specific surface area, diameter of pores, etc.), and very little upon the technique of brewing, being about two parts to one part coffee for a coarse grind.
—this quantity of fixed water, on the other hand, decreases as a proportion of the volume of liquid (water) taken, as the latter increases. Comparable changes can be found by use of a degree of grind more fine or more coarse.

3.3.2. Concentrations of Soluble Solids

The curves in Fig. 9 clearly show a progressive lowering of the concentration of soluble solids in the extract, corresponding to an increase in the volume of the extracting liquid (water) for a given quantity of ground coffee. They equally show that two of the techniques utilised (continuous water circulation and filtration) do not allow a reduction of this volume below a certain limit. Though the slope of the curves may not be absolutely identical, the changes in soluble solids concentrations are very similar for the different types of equipment employed. Filtration by gravity, by reason of the particularly long duration of brewing, and the vaporisation technique, on account of optimum contact between the roast coffee and the extracting liquid, provide the most concentrated extracts. Filtration under pressure, in contrast, provides relatively dilute extracts, by reason principally of its very short (possibly too short) period of extraction.

For the two procedures for which the duration can be altered by the

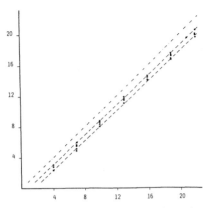

Fig. 8. Water retention by the extracted coffee: relationship between the relative proportions of liquid brew to coffee (w/w) and the relative proportions of water to coffee (w/w). Top line = zero retention. Medium grind.

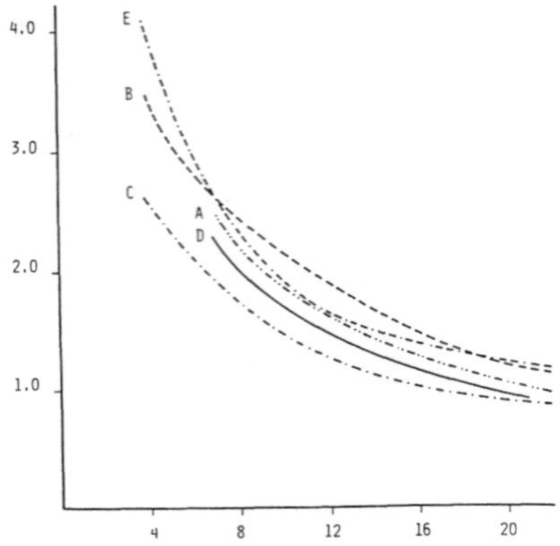

Fig. 9. Comparison of brewing techniques, I: relationship between the amount of solids in the brew (%) and the relative proportions of water to coffee (w/w). A, Continuous water circulation ('Cona' or vacuum maker); B, continuous filtration by gravity (filter); C, filtration by gravity under pressure; D, filtration with recirculation percolator); E, vaporisation under pressure (espresso).

consumer (continuous water circulation, and filtration with recirculation or 'percolator'), the length of time used in our trials had been chosen taking mainly into account organoleptic considerations. This period of time, which was about 6 min for the highest amount of water used, appeared too short to obtain a sufficiently concentrated extract.

3.3.3. Extraction of the Ground Coffee

The extraction yield applicable to the raw material depends in a logical manner on the soluble solids concentration, since the weight of extract obtained is a direct function of one of the parameters varied (i.e. water/coffee ratio). The results on the subject of soluble solids concentration are therefore valuable for evaluating extraction yields obtained from the raw material, the values of which for a medium grind are collected in Fig. 10. Independent of the method employed in each of the types of equipment, it is the length of processing which primarily influences the solubilisation of the solid constituents of the coffee. In order to show this influence, it is sufficient to compare the curves which refer respectively to the techniques of filtration by gravity, one that is limited to a simple lixiviation (B) (percolation); and that which proceeds (in a short time) under a slight overpressure (C). The relatively small difference between the extractions obtained by using a restricted amount of water is accentuated with an increase in the proportion of water to coffee.

In order to determine in practice the role played by the degree of grind of the coffee, the average diameter of the particles was varied between 650 and 1200 μm (with satisfactory coefficients of variation from Rosin–Rammler diagrams). The curves of Fig. 11 show clearly the effects of grinding on the passage of the soluble solid constituents into solution. For the equipment in which the total length of brewing is long, but where the actual time of contact between the liquid and solid phases is relatively short (filtration by gravity), these effects are relatively marked. For apparatus where the time of contact is prolonged in a repetitive manner (filtration with recirculation, 'percolator'), the variations in the degree of extraction or exhaustion at different grinds are rather limited and seem practically independent of the relative proportions used in the brewing.

3.3.4. Aromatic Content of Brews

By applying an efficient evaporation in suitable apparatus to each of the brews examined, it is possible to separate the essential volatile components, which are then recovered by condensation. The total content of these substances can be determined by a simple measurement of their absorption in the UV region in the corresponding aqeuous condensate. This technique provides only a purely indicative idea of the relative richness of the brews in aromatic content. The results of these spectrophotometric determinations have been collected together in Fig. 12, where they have been expressed as a proportion of the initial aromatic content of the ground coffee.

In spite of the importance of the quantity of liquid taken for the extraction in order to obtain higher proportions, it is not possible to achieve total solubilisation of the volatile substances present in the raw material. This incomplete transmission seems to correspond to a considerable retention of these substances in the spent grounds (lipid fraction[102]) or by the portion of extract absorbed by the latter. The brewing techniques, in which the preparation demands a prolonged time of contact between the material to be extracted and the liquid of extraction, procure extracts particularly poor in aromatic principles. These techniques, continuous water circulation (A) and filtration with recirculation (D)—

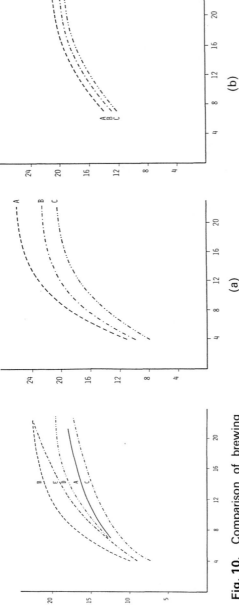

Fig. 10. Comparison of brewing techniques, II: relationship between the extraction yield (%) and the relative proportions of water to coffee (w/w). A, continuous water circulation ('Cona' or vacuum maker); B, continuous filtration by gravity (filter); C, filtration by gravity under pressure; D, filtration with recirculation (percolator); E, vaporisation under pressure (household espresso).

Fig. 11. Relationship between the extraction yield (%) and the relative proportions of water to coffee (w/w). A, Fine grind; B, medium grind; C, coarse grind. (a) Influence of the degree of grinding (filtration by gravity). (b) Influence of the degree of grinding (filtration with recirculation).

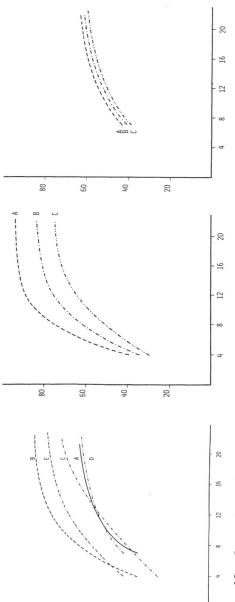

Fig. 12. Comparison of brewing techniques, III: relationship between the content in volatile components of the brew as a function of the starting material (%) and the relative proportions of water to coffee (w/w). A, continuous water circulation ('Cona' or vacuum maker); B, continuous filtration by gravity (filter); C, filtration by gravity under pressure; D, filtration with recirculation (percolator); E, vaporisation under pressure (household espresso).

Fig. 13. Relationship between the content in volatile components of the brew as a function of the starting material (%) and the relative proportions of water to coffee (w/w) A, Fine grind; B, medium grind; C, coarse grind. (a) Influence of the degree of grinding (filtration by gravity—filter). (b) Influence of the degree of grinding (filtration with recirculation—percolator).

especially the former—are very probably responsible for vaporisation loss of an important part of the volatile components.

As would appear logical, the degree of grind applied has an influence on the passage of aromatic constituents into solution comparable to that already mentioned for the solubilisation of the soluble solids (cf. curves of Fig. 13). This influence depends also upon the technique applied for the brewing, since marked differences are again evident between filtration by gravity and filtration with recirculation. The losses by vaporisation mentioned above in this latter technique are again manifested, irrespective of the degree of grind used.

3.3.5. Solubilisation of Individual Constituents

The principal individual constituents of a coffee brew have been determined on a large number of representative samples, which were previously freeze-dried. In order to allow comparison of all the results, the combined results obtained (using a fine grind), from all the types of equipment, are given in Fig. 14; those referring to brews from a coarser grind are to be found in Fig. 15.

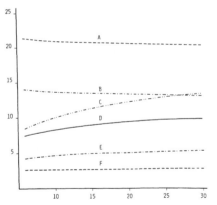 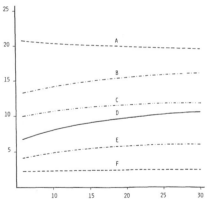

Fig. 14. Solubilisation of the solid constituents (fine grind): relationship between the amount of each compound, expressed as a percentage of the original dry material, and the extraction yield of the brewed coffee (%). A, Total reducing sugars; B, total ashes; C, presumed proteins; D, chlorogenic acids; E, caffeine; F, direct reducing sugars.

Fig. 15. Solubilisation of the solid constituents (coarse grind): relationship between the amount of each compound, expressed as a percentage of the dry material, and the extraction yield of the brewed coffee (%). A, Total reducing sugars; B, total ashes; C, presumed proteins; D, chlorogenic acids; E, caffeine; F, direct reducing sugars.

As the slopes of the different curves are generally similar and the standard deviation of the values taken sometimes reaches $\pm 5\%$, it is important to envisage above all the general trend of the information obtained by these analytical determinations. This trend is, moreover, modified when the importance of the weight of constituents present in the brew is examined, and not their content expressed as a function of the total solid matter.

Certain constituents analysed show themselves as easily extractable substances, since their content shows a slight diminution with increase of extraction yield. These are, in particular, reducing groups, which are liberated by acid hydrolysis and designated as 'total sugars' (A) in Figs 14 and 15; and the total inorganic content (B), but only in the case of a fine grind. All the other components, on the contrary, are seen to increase their content moderately as a function of increase in extraction yield, behaving thus as less soluble substances, at least under the conditions used with these brews. The use of coarser grind is responsible for little marked modifications, apart from a change in the behaviour of total ash already noted. We should, however, note a perceptible diminution of the content of reducing sugars, partially compensated by an increase of 'protein', which may be a mixture of nitrogenous components with the exception of caffeine.

3.3.6. Sensory Evaluation

Graphical representation. The sensory evaluation of each of the brews submitted to the panel of experts can be represented graphically in the form of a relative taste profile, as expressed by ratioing the values of particular attributes to those in the same reference brew. As a demonstration, the aroma and taste characteristics of a brew prepared starting with a fine grind by filtration under pressure have been set out in Fig. 16. The relative length of the vectors shown corresponds, for each of the aromatic or taste characteristics considered, to its difference in intensity by ratio (positive/negative) to that of the reference.

In the first case envisaged, which involves a low extraction yield obtained by using a limited volume of liquid, differences manifest themselves in a negative manner (with the exception of the 'cereal' note generally little appreciated by consumers). When one increases the volume of extracting liquid, and thus the extraction yield, one notes a marked diminution in the intensity of the expressed differences; without, however, causing a reversal of preference. It is only on attaining the maximum volume of liquid usable by the equipment that a reversal is achieved which

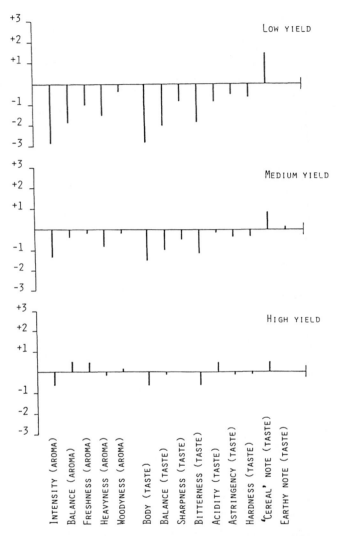

Fig. 16. Relative flavour profile of the brews prepared by filtration under pressure (fine grind).

manifests itself, in particular, in the balance and freshness impressions of the aromatic character as well as in the natural acidity of the brew. One notes further a persistence of the 'cereal' note, typical of this type of equipment.

Canonical analysis. The evaluations given by the different tasters of the brews which have been successively submitted to them have been processed with the aid of a canonical analysis, designed to position the 14 sensory criteria enumerated in the experimental work in a multi-dimensional space. In order to simplify the interpretation of this complex structure, it is preferable to project it following two principal axes; nearly 80% of the information obtained by the mathematical analysis is represented in this bidimensional projection (Fig. 17). The length of each of these vectors corresponding to each of the sensory attributes is proportional to the discrimination of this attribute. Certain of them, such as the intensity (numbered 1) and body (numbered 6) of the brew, are therefore particularly sensed by all the judges participating in the evaluation. Some of the others, such as the 'woodiness' of the aroma or the 'earthy' character of the taste, seemed not to be perceived in a significant manner in the brews tested.

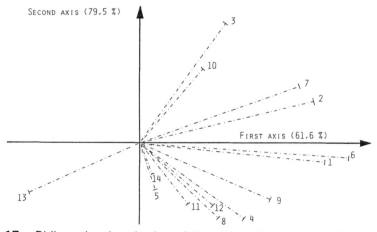

Fig. 17. Bidimensional projection of the principal sensorial attributes of a coffee brew. *Aromatic characteristics:* 1, intensity; 2, balance; 3, freshness; 4, heaviness; 5, woodiness. *Taste characteristics:* 6, body; 7, balance; 8, sharpness (negative attribute); 9, bitterness (positive attribute); 10, acidity (positive attribute); 11, astringency (negative attribute); 12, hardness; 13, 'cereal' note; 14, earthy note.

The attributes referring to the intensity and balance of a brew, i.e. those designated in Fig. 17 by the numbers 1, 2, 6 and 7, are those that contribute the most to the first principal axis. One can therefore consider that this latter attribute expresses, in a positive sense, the fullness of the natural characteristics of coffee submitted to extraction. All the attributes contributed to a greater or lesser degree to this fullness, with the exception of the 'cereal' note, which manifests itself in a diametrically opposed sense. The second principal axis is of a bipolar nature, and its positive direction coincides principally with the elements of freshness and of 'fineness' present in the cup. In opposition, its negative direction is above all influenced by the characteristics of bitterness, acidity and harshness, which arise in general from a higher degree of roasting.

From their position in the projection, it is apparent that certain of the attributes are intimately associated in the evaluation by the tasters. This is the case notably with the intensity of aromatic character which is related to the full-bodied character of taste in the same way that the freshness of the aroma corresponds to the acidity of the product examined. The attributes of heaviness of the aroma on the one hand and astringency, bitterness, sharpness and harshness of the taste on the other constitute equally a family of characteristics strongly correlated with each other.

By applying to the evaluation of different judges a discriminatory analysis by successive stages, it is possible to disengage, among the 14 attributes used, those that allow, taken together, an essentially complete description of each of the brews. These attributes are five in number, thus by order of introduction in the mathematical model: character 6, body or its fullness; character 7, balance of the taste; character 10, natural acidity; character 1, intensity of the aroma character; and character 2, balance of the aroma. Among these fundamental attributes, it is the acidity that procures the most basic discrimination, without reference to any of the other attributes used.

Similarly to that which has been demonstrated for individual sensory characteristics, a distribution within a bidimensional projection of the different products that have been submitted to the judgement of tastes is equally possible. This positioning is represented in Fig. 18, which takes into account the degree of grind. The code used to designate the products examined can be summarised as follows: the first capital letter refers to the technique of brewing which is described in the caption of the figure, and the second letter refers uniquely to the extraction yield ('L', significantly low, 'M' medium, and 'H' high).

The position in the projection plan of the brews examined by reference

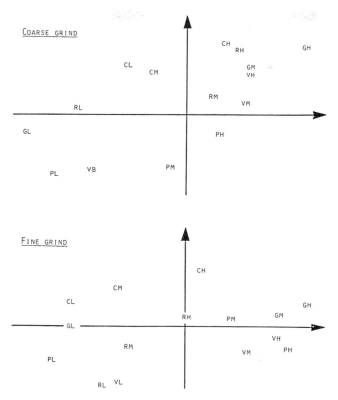

Fig. 18. Bidimensional projection of the various brews evaluated organoleptically. *First capital letter:* C, continuous circulation ('Cona' or vacuum maker); G, filtration by gravity (filter); P, filtration under pressure; R, filtration with recirculation (percolator); V, vaporisation under pressure (household espresso). *Second capital letter:* L, low yield; M, medium yield; H, high yield.

to the two principal axes and especially by reference to the projection of the sensory attributes can provide very valuable indications of the dominant properties. The origin between the two axes representing average quality of a brew as a function of the parameters used is derived from the sample prepared by filtration with recirculation, with a maximum volume of liquid and a fine grind, which approaches the medium position most closely.

For a given brewing technique, increase in volume of the liquid, that is to say increase of extraction yield, causes a displacement of the position on the plan of the corresponding brew, parallel to the positive direction of the first principal axis. This displacement is equivalent to a reinforcement of the fullness and balance of the aromatic and taste properties of this brew. In most cases, a movement in the positive sense of the second principal axis can equally be established, which signifies an improvement in the quality and freshness of the brew considered. Change of grind degree, coarser to more fine, generally finds expression in a movement in the negative sense along this second axis, thus indicating a perceptible diminution of these two essential characteristics.

When one compares organoleptically the particular merits of each of these brewing techniques, one establishes that the continuous water circulation method (A) produces brews of highest quality and freshness but singularly lacking in the full body characteristics indispensable to the general quality of the brew. Vaporisation under pressure (E) for its part furnishes brews possessing markedly the innate characteristics of the coffee, where the bitterness and harshness are particularly dominant but in which the fineness and freshness already mentioned are unfortunately absent. Filtration under pressure (C) and filtration with recirculation (D) are responsible for brews of medium quality without any character truly dominant but clearly inferior in quality to those coming from the other types of equipment compared. It is therefore filtration by gravity (B) which gives extracts the most balance, allied with natural acidity and bitterness, aromatic intensity and fullness of taste.

Acknowledgements

The work for which the results have been presented in the preceding pages would not have been achieved without the assistance of M. Luc Vuataz, who has carried out all the mathematical parts of the study, and Mme Monika Corbaz-Sticht, who has accomplished the essential work of preparation and analysis of the brews. They must be thanked here for their efficiency and diligence. Our thanks are addressed also to the panel of tasters for their regular and enthusiastic participation in the evaluation sessions.

4. GENERAL CONCLUSIONS

The bibliographical research that we have conducted during the preparation of this chapter has fully taken into account the very large number of scientific publications, appearing mainly during the last 20 years, which refer to coffee brews and beverages and their preparation. A résumé of these publications allows us to draw the following conclusions.

The most detailed and complete studies have been concerned, in the first place, with the different techniques of brewing and the apparatus available for their application. The physical phenomena that cause the passage of the soluble constituents, considered in their entirety, from the initial solid phase to the absorbed liquid phase, then their diffusion to the external liquid phase, have been established experimentally in a precise manner. As the theoretical hypotheses concerning the successive stages have been developed, some very good correlations have been established between the values obtained by mathematical modelling and those procured by practical trials.

If one moves from the 'solid–liquid' extraction, envisaged as an overall operation, to considerations of the particular mechanisms responsible for the transmission of individual constituents from the raw material to the consumable brew, one notes that important gaps exist in the available knowledge on this subject. Though the behaviour of a well-defined substance such as caffeine can be determined with certainty, so many of those constituents of a more complex nature, such as the carbohydrates and non-caffeinic nitrogenous components, are difficult to specify. For these types of substance the effects attributable to the solubilisation are difficult to distinguish from the modifications ascribable to the parameters exercised during the extraction, particularly the temperature. This is a field of research that has been insufficiently explored, and should be studied in the years to come by the analytical experts.

For a given brew, arising from well defined and fully reproducible parameters, the establishment of the principal and secondary sensory characteristics has become relatively easy when the recently developed evaluation techniques are applied. The specific attributes of the aroma and taste of the coffee brew can be evaluated qualitatively and quantitatively by a group of trained tasters by comparison with a reference product or in an entirely independent manner.

Some incontestable difficulties on the contrary manifest themselves when trying to correlate the relative intensity of one of these sensory attributes with the importance of one or several chemical constituents. The existence of simple relationships between the perception of certain aromatic characteristics and the content of well defined volatile principles has already been noted. But fundamental research over several years is still necessary for indisputable correspondence to be established between organoleptic properties and the chemical composition of a coffee brew.

REFERENCES

1. Pictet, G. and Vuataz, L., *Proc. 8th Coll. ASIC*, 1977, 261–9.
2. Brünische-Olsen, H., *Solid Liquid Extraction*, NYT Nordisk Forlag, Arnold Busck, Copenhagen, 1965.
3. Long, V. D., *J. Fd Technol. (UK)*, 1977, **12** (5), 459–72.
4. Voilley, A., Sauvageot, D. and Durand, D., *Proc. 8th Coll. ASIC*, 1977, 251–60.
5. Bischel, B., *Fd Chem.*, 1979, **4**, 53–62.
6. Dousse, R., Dissertation, Polytechnicum de Zürich, 1978.
7. Leniger, H. A. and Beveloo, W. A., *Food Process Engineering*, D. Reidel, Boston (USA), 1975.
8. Spanninks, J. A. M., Dissertation, Agricultural University of the Netherlands, Wageningen, 1979.
9. Schwartzberg, H. G., *Chem. Engng Progr. (A.I.Chem.E.)*, April 1980, **21**, 67–85.
10. Schwartzberg, H. G. and Chao, R. G., Communication to Annual Meeting of Food Technologists, Atlanta (USA), 7–10 June 1981.
11. Clo, G., Dissertation, ENSBANA, Université de Dijon, 1981.
12. Caille, J., Mémoire ENITA, Dijon, 1975.
13. Loncin, M., *Génie industriel alimentaire—Aspects fondamentaux*, Masson, Paris, 1976.
14. MacCabe, W. L. and Smith, J. C., *Unit Operations of Chemical Engineering*, McGraw-Hill, New York, 1956.
15. Geankoplis, C. J., *Transport Processes and Unit Operations*, Allyn & Bacon (USA), 1978.
16. Thaler, H., *Deutsche Lebensm. Rundschau*, 1955, **51**, 283–6.
17. Sivetz, M. and Foote, H. E., *Coffee Processing Technology*, AVI, Westport, Conn., 1963.
18. Sivetz, M. and Desrosier, N. W., *Coffee Technology*, AVI, Westport, Conn., 1979.
19. Streuli, H., in *Handbuch der Lebensmittelchemie*, Band VI, Ed. J. Schormüller, Springer, Berlin, 1970.
20. Maier, H. G., *Kaffee*, Paul Parey, Hamburg, 1981.
21. Heiss, R. and Radtke, A., *Kaffee Tee-Markt*, 1969, **19** (8), 35–43.
22. Anon., *Kaffee Tee-Markt*, 1975, **25** (19), 18.
23. Coste, R., *Les caféiers et les cafés dans le monde, 1ᵉ partie*, Larose éditeur, Paris, 1959.
24. Natarajan, C. P., Shivashanhar, A., Balachandra, A. and Bhatia, D. S., *Indian Coffee*, 1965, **29**, 5–7.
25. Anon., *Kaffee Tee-Markt*, 1971, **21** (10), 3, and **21** (20), 3, 22.
26. Anon., *Tea Coffee Trade J.*, 1974, **146** (4), 20–1, 23.
27. Coffee Brewing Institute, *Coffee Tea Ind.*, 1957, **80** (4), 44–78.
28. Coste, R., Le caféier, in *Techniques agricoles et productions tropicales*, Maisonneuve et Larose, Paris, 1968.
29. Bürgin, E., *Proc. 4th Coll. ASIC*, 1969, 63–74.
30. Biggers, R. E., Hilton, J. J. and Gianturco, M. A., *J. Chrom. Sci.*, 1969, **7**, 453–72.

31. Chassevent, F., Vincent, J. C., Hahn, D., Pougnaud, S. and Wilbaux, R., *Proc. 4th Coll. ASIC*, 1969, 179–85.
32. Van der Stegen, G. H. D., *Fd Chem.*, 1979, **4** (1), 23–31.
33. Tomoda, G., Matsuyama, J., Nagano, A., Namatame, M. and Morita, Y., *J. Japan. Soc. Fd Sci Technol.*, 1980, **27** (2), 68–74.
34. Heiss, R. and Radtke, R., *Kaffee Tee-Markt*, 1969, **19** (5), 5–12.
35. Clydesdale, F. M. and Francis, F. J., *Fd Prod. Devel.*, 1971, **5** (2), 67–78.
36. Lerici, C. R., Lercker, G., Minguzzi, A. and Matassa, P., *Industrie delle Bevande*, 1980, **9**, 232–8.
37. Lerici, C. R., Dalla Rosa, M., Magnanini, E. and Fini, P., *Industrie delle Bevande*, 1980, **9**, 375–81.
38. Pictet, G. and Rehacek, J., *Proc. 10th Coll. ASIC*, 1982, 219–33.
39. Arjona, J. L., Rios, G., Gibert, H., Vincent, J. C. and Roche, G., *Café Cacao Thé*, 1977, **21**, 263–72.
40. Arjona, J. L., Rios, G. and Gibert, H., *Lebensm. Wiss. u. Technol.*, 1980, **13**, 285–90.
41. Maier, H. G., *Proc. 11th Coll. ASIC*, 1985, 291–6.
42. Kazi, T. and Clifford, M. N., *Proc. 11th Coll. ASIC*, 1985, 297–308.
43. Gardner, D. G., *Fd Res.*, 1958, **23**, 76–84.
44. Punnett, P. D., *Tea Coffee Trade J.*, 1963, February, 26–8.
45. Pangbourn, R. M., *Proc. 8th Coll. ASIC*, 1977, 249–50.
46. Pangbourn, R. M., *Lebensm. Wiss. Technol.*, 1982, **15**, 161–8.
47. Lockhart, E. E., Tucker, C. L. and Merritt, M. C., *Fd Res.*, 1955, **20**, 598–605.
48. Glatzel, H., *Z. Lebensm. Unters. Forsch.*, 1962, **119**, 26–30.
49. Pangbourn, R. M., Tabue, I. M. and Little, A. C., *J. Fd Sci.*, 1971, **36**, 355–62.
50. Eyer, E., *Ernährung-Umschau*, 1975, **22** (6), 167–9.
51. Punnett, P. D., *Tea Coffee Trade J.*, 1962, May, 32–4.
52. Voilley, A. and Simatos, D., *Abstracts of the 5th International Congress on Food Science and Technology, Kyoto, 1978*, 122.
53. Voilley, A. and Simatos, D., *J. Fd Proc. Engng*, 1980, **3**, 184–98.
54. Voilley, A., Sauvageot, F., Simatos, D. and Wojcik, G., *J. Fd Proc. Preserv.*, 1981, **5** (3), 135–43.
55. Lockhart, E. E., *Fd Res.*, 1959, **24**, 91–6.
56. Clo, G. and Voilley, A., *Lebensm. Wiss. Technol.*, 1983, **16**, 39–41.
57. Clo, G., Mémoire, D. E. A., ENSBANA, Université de Dijon, 1979.
58. Ndjouenkeu, R., Dissertation, ENSBANA, Université de Dijon, 1983.
59. Illy, E. and Ruzzier, L., *Café Cacao Thé*, 1963, **7**, 385–94.
60. Meulman, H. C., *Voeding* (Dutch), 1963, **24**, 696–706.
61. Barbara, C. E., Capella, P. and Carnacini, A., *Industrie agrarie*, 1968, **6** (11), 561–70.
62. Adinolfi, J., *World Coffee and Tea*, 1974, September, 20, 24, 26, 27.
63. Lindner, M. W., *Deutsche Lebensm. Rundschau*, 1951, **47**, 204–5.
64. Voilley, A., Sauvageot, F. and Pierret, P., *Proc. 9th Coll. ASIC*, 1980, 287–94.
65. Kovacs, A. C. and Wolf, H. O., *Kaffee Tee-Markt*, 1963, **8** (4), 6–12.
66. Maier, H. G. and Sander, B., *Chem. Mikrobiol. Technol. Lebensm.*, 1975, **3**, 164–8.

67. Sakane, Y. et al., *J. Japan. Soc. Fd Sci. Technol.*, 1984, **31** (2), 66–71.
68. Hyvönen, L. et al., *J. Fd Sci.*, 1978, **43**, 1577–9.
69. Hughes, E. B. and Smith, R. F., *J. Soc. Chem. Ind.*, 1949, **68**, 322–7.
70. Segall, S. and Proctor, B. E., *Fd Technol.*, 1959, **13**, 266–9, 383–4, 679–83.
71. Maier, H. G., Engelhardt, U. H. and Scholze, A., *Deutsche Lebensm. Rundschau*, 1984, **80** (9), 265–8.
72. Segall, S., Silber, C. and Bacino, C., *Fd Technol.*, 1970, **24**, 1242–6.
73. Belitz, H. D., *Proc. 7th Coll. ASIC*, 1975, 243–52.
74. Ferreira, L. A. B. et al., *Proc. 5th Coll. ASIC*, 1971, 79–84.
75. Kulaba, G. W., MSc Thesis, University of Nairobi, Kenya, 1978.
76. Van Roekel, J., *Proc. 7th Coll. ASIC*, 1975, 259–64.
77. Maier, H. G., Balcke, C. and Thies, F. C., *Deutsche Lebensm. Rundschau*, 1983, **37**, 81.
78. Belitz, H. D. et al., *Chem. Ind.*, 1983, 23–6.
79. Lea, A. G. H. and Arnold, G. M., *J. Sci. Fd Agric.*, 1978, **29**, 478–83.
80. Maier, H. G., Balcke, C. and Thies, F. C., *Deutsche Lebensm. Rundschau*, 1984, **80**, 367–9.
81. Amerine, M. A., Pangbourn, R. M. and Roessler, E. B., *Principles of Sensory Evaluation of Foods*, Academic Press, New York, 1965.
82. Ordynski, G. Z., *Ernährungwiss.*, 1965, **6**, 203–6.
83. Ariga, T. and Asao, Y., *Agric. Biol. Chem.*, 1981, **45**, 2709–12.
84. Clifford, M. N. and Ohiokpehai, O., *Anal. Proc.*, 1983, **20**, 83–6.
85. Smith, A., Presentation to the Society of Chemical Industry, London, 1983.
86. Arnold, R. A. and Noble, A. C., *Amer. J. Enol. Viticulture*, 1978, **32**, 5–13.
87. Punnett, P. D., *Tea Coffee Trade J.*, 1962, June, 26–30.
88. Amorim, H. V., Malavola, E., Teixeira, A. A., Cruz, V. F., Melo, M., Guercio, M. A., Fossa, E., Breviglieri, O., Ferrari, S. E. and Silva, D. M., *Proc. 6th Coll. ASIC*, 1973, 113–27.
89. Tassan, C. G. and Russel, G. F., *J. Fd Sci.*, 1974, **39**, 64–8.
90. Pangbourn, R. M., Gibbs, Z. M. and Tassan, C. G., *J. Texture Studies*, 1978, **9**, 415–36.
91. Vitzthum, O. G., in *Kaffee und Coffein*, Ed. O. Eichler, Springer, Berlin, 1976.
92. Navellier, P., Brunin, R., Chassevent, F. and Isaac, A., *Ann. Fals. Fraudes*, 1960, **53**, (618), 326–36.
93. Thaler, H., *Deutsche Lebensm. Rundschau*, 1957, **53**, 49–51.
94. Lockhart, E. E., *Tea Coffee Trade J.*, 1957, **113** (1), 12–13 and 45–48.
95. Lockhart, E. E., *Coffee Tea Ind.*, 1959, **82** (9), 16–17.
96. Wurziger, J., *Proc. 3rd Coll. ASIC*, 1967, 299–312.
97. Ara, V. and Thaler, H., *Z. Lebensm. Unters. Forsch.*, 1976, **161**, 143–50.
98. Navellier, P., in *Les caféiers et les cafés dans le monde*, Ed. R. Coste, Larose, Paris, 1959.
99. Lockhart, E. E., Coffee Brewing Institute Publication No. 25, New York City, 1957.
100. Reichstein, T. H. and Staudinger, H., *Ciba Zeitschrift (Basel)*, 1951, **11** (127), 4692–4.
101. Weurman, C., *Café Cacao Thé*, 1963, **7**, 341–6.
102. Streuli, H., in *Aroma- und Geschmackstoffe in Lebensmitteln*, Eds J. Solms and H. Neukom, Forster, Zurich, 1967.

103. Vitzthum, O. G., Barthels, M. and Kwasny, H., *Z. Lebensm. Unters. Forsch.*, 1974, **154**, 135–40.
104. Merritt, M. C. and Proctor, B. E., *Fd Res.*, 1959, **24** (6), 735–43.
105. Thomczik, C., *Lebensm.-Ind.*, 1979, **26** (12), 561–3.
106. Ndjouenkeu, R., Clo, G. and Voilley, A., *Science des Aliments*, 1981, **1** (3), 365–75.
107. Werner, H. and Kohley, M., *Kaffee Tee-Markt*, 1965, **15** (1), 5–11, (2), 6–12, (3), 6–12, (4), 6–10, (5), 5–10.
108. Takahashi, K., Kondo, Y. and Sawano, T., *J. Japan. Soc. Fd Sci. Technol.*, 1979, **26** (8), 360–1.
109. Horman, I. and Viani, R., *Proc. 5th Coll. ASIC*, 1971, 102–11.
110. Van der Stegen, G. H. D. and Van Duijn, J., *Proc. 9th Coll. ASIC*, 1980, 107–12.
111. Kampmann, B. and Maier, H. G., *Z. Lebensm. Unters. Forsch.*, 1982, **175**, 333–6.
112. Maier, H. G. and Grimsehl, A., *Kaffee Tee-Markt*, 1982, **32** (23), 3–5.
113. Zawadska, J. and Leki, B., *Przemysl Spozywczy (Poland)*, 1968, **22** (10), 461–2.
114. Jakober, P. and Staub, M., *Mitt. Lebensm. Unters. Hyg.*, 1963, **54**, 26–34.
115. Clarke, R. J. and Walker, L. J., *J. Sci. Fd Agric.*, 1974, **25**, 1389–1404.
116. Clarke, R. J. and Walker, L. J., *Proc. 7th Coll. ASIC*, 1975, 159–64.
117. Fragoso, M. A. C., Feirrera, L. A. B. and Peralta, M. F., *Proc. 5th Coll. ASIC*, 1971, 70–8.
118. Lara, W. H., De Toledo, M. and Takahashi, M. Y., *Revista do Instituto Adolfo Lutz (Brazil)*, 1975–1976, **35–36**, 17–22.
119. Stenstroem, T. and Vahter, M., *Var. Foeda (Sweden)*, 1975, **27** (3), 150–6.
120. Teply, L. J. and Prier, R. F., *J. Agric. Fd Chem.*, 1957, **5**, 375–7.
121. Kwasny, H., *Lebensm. Gerichtl. Chemie*, 1978, **32** (2), 36–8.
122. Thaler, H. and Gaigl, R., *Z. Lebensm. Unters. Forsch.*, 1963, **120**, 449–54.
123. Natarajan, C. P. *et al.*, Fd Sci. (Mysore), 1958, **7**, 53–5.
124. Barbetti, P. and Chiappini, I., *Industrie delle Bevande*, 1977, **6** (6), 97–9.
125. Nakabayashi, T., *J. Japan. Soc. Fd Sci. Technol.*, 1984, **31** (6), 421–2.
126. Lehmann, G. and Moran, M., *Kaffee Tee-Markt*, 1973, **23** (9), 10–11.
127. Santanilla, J. D., Dissertation, Universität Munster (D), 1977.
128. Wyler, O. and Högl, O., *Mitt. Lebensm. Unters. Hyg.*, 1948, **39**, 351–71.
129. Wurziger, J., *Kaffee Tee-Markt*, 1956, **6** (21), 5–6, (22), 4–6, (23), 4–8.
130. Merritt, M. C. and Proctor, B. E., *Fd Res.*, 1959, **24**, 672–80.
131. Pekkarinen, L. and Porkka, E., *Z. Lebensm. Unters. Forsch.*, 1963, **120**, 20–5.
132. Hohlfeld, W., *Z. Lebensm. Unters. Forsch.*, 1959, **110**, 129–30.
133. Sarudi, I. J. and Siska, E., *Deutsche Lebensm. Rundschau*, 1970, **66** (4), 127–30.
134. Czechowska, M., Golobiowsky, T., Pieczonka, W. and Solarz, A., *Die Lebensmittel-Industrie*, 1976, **23** (H.6), 258–60.
135. Riedel, L., *Chem. Mikrobiol. Technol. Lebensm.*, 1974, **4**, 108–12.
136. Reymond, D., Pictet, G. and Egli, R. H., *Proc. 2nd Coll. ASIC*, 1965, 150–60.
137. Haevecker, U., *Proc. 4th Coll. ASIC*, 1969, 160–5.

138. Borker, E. and Sloman, K., *J. Ass. Off. Agric. Chem.*, 1965, **48** (4), 705–9.
139. Haüsermann, H. and Brandenberger, H., *Z. Lebensm. Unters. Forsch.*, 1961, **115**, 516–27.
140. Navellier, P., Brunin, R. and Sartre, J. P., *Proc. 2nd Coll. ASIC*, 1965, 49–54.
141. Timmell, T. E., Glaudemans, C. P. and Curie, A. L., *Anal. Chem.*, 1956, **28** (12), 1916–20.
142. Pictet, G., *Proc. 7th Coll. ASIC*, 1975, 189–200.

Chapter 9

Waste Products

M. R. ADAMS[1] and J. DOUGAN[2]

[1] *Department of Microbiology, University of Surrey, Guildford, UK*
[2] *Tropical Development and Research Institute, London, UK*

1. PRIMARY PROCESSING: THE PRODUCTION OF GREEN COFFEE

The green coffee bean of international commerce constitutes only 50–55% of the dry matter of the ripe cherry (18% on a fresh weight basis). The remaining material is diverted into a variety of low- or negative-value by-products depending on the processing technique used. Here we describe these by-products and the various schemes that have been proposed for their utilisation or disposal, and attempt a critical evaluation of their feasibility. Aspects of this subject have been discussed previously by several authors.[1-6]

1.1. Dry or Natural Processing

Dry processing is the simplest technique for processing coffee cherries, both technologically and with regard to the waste disposal problem it creates. It does, however, produce coffee inferior in quality to the wet processed product. Most robusta beans, currently comprising about 30% of world production, and Brazilian arabica beans, roughly 30–50% of world arabica production, are processed in this way.[7-10]

The coffee cherries are left on the tree until overripe. After harvesting they are further dried to about 10–11% moisture and the layers of material overlaying the beans, known as the husk, are removed simultaneously in a dehulling machine. Proximate analyses of robusta and arabica husks, the sole by-product from dry processing, are presented in Table 1.

Table 1
Yield of solid wastes from processing 1 kg fresh coffee cherry

Analysis (%)	Wet processing				Parchment (35–61 g) Dry basis[b]	Dry processing Husks (180 g)	
	Pulp (400–432 g)						
	Fresh[a]	Dried[a]	Naturally[a] fermented and dried	Dry basis[b]		C. robusta[c] Dry basis	C. arabica[c] Dry basis
Moisture	76·7	12·6	7·9	0	0	0	0
Dry matter	23·3	87·4	92·1	100	100	100	100
Ether extract	0·48	2·5	2·6	1–2	0·5	2·0	1·7
Crude fibre	3·4	21·0	20·8	12–20	50	27·6	13·2
Crude protein	2·1	11·2	10·7	4–12	1–2·5	9·2	11·3
Ash	1·5	8·3	8·8	6–10	0·5–1	3·3	6·8
Total pectic substances[d]	6·5	—	—	6	—	6·5	—
Reducing sugar[d]	12·4	—	—	—	—	12·4	—
Non-reducing sugar[d]	2·0	—	—	—	—	2·0	—
Total sugar[d]	14·4	—	—	14·0	—	14·4	—

[a] Bressani et al.[11]
[b] Rolz.[12]
[c] Tango[2]
[d] Dry basis.

1.1.1. The Use of Husks as Fuel

Dried coffee husks are a useful source of cheap fuel with a calorific value of around $14\,MJ\,kg^{-1}$.[13] They are used extensively for this purpose in coffee drying, and since any excess may be burnt they pose no serious disposal problem. Conversion of surplus husks to charcoal, in which the energy is more concentrated (calorific value $29\cdot4\,MJ\,kg^{-1}$), would reduce transport costs and may offer some positive financial return (see section 1.2.10).[14] Gasification and hydrogenation techniques for the thermal conversion of relatively dry plant material are under development in many countries and could be applied to coffee husks in the future.[15]

1.1.2. The Use of Husks as Animal Feed

Rogerson assessed the nutritive value of coffee husks and concluded that they could not be regarded as either a useful basic feed or as a concentrate supplement since the digestible protein content is very low and the starch equivalent value did not exceed that of fairly poor quality hay.[16] They do have the virtue of being readily available at low cost during the season, and subsequent trials in East Africa have shown that husks can be fed to steers up to levels of 20% of the total ration without affecting food intake, weight gain or feed conversion.[17,18] Similar conclusions have been reached in studies with dairy cattle in India which also found that a 20% inclusion rate in the ration of dairy cows had no significant effect on the quantity and quality of the milk yield[19,20] (see also section 1.2.12).

1.2. Wet processing

Wet processing, described in detail in Chapter 1 in this volume, consists of the mechanical removal of the pulp (the exocarp and most of the mesocarp of the fruit) in the presence of water. The beans obtained in this way are surrounded by a layer of mucilaginous material 0·5–2 mm thick which must be removed to facilitate drying. This is normally achieved through a natural fermentation process in which micro-organisms act as the principal agents in solubilising the mucilage pectin.[21] After fermentation the clean bean is dried to about 12% moisture and the parchment (endocarp) and silverskin (testa) are removed by a hulling machine.

This technique, which accounts for about half the world's coffee production[22] is acknowledged to produce the finest quality product. To be offset against this is the fact that it is more complex than the dry process, requires a greater capital investment, more power and greater

Table 2
Pollution load of liquid wastes from coffee processing

	Pulping waters	Fermentation waters
Chemical oxygen demand, COD (ppm)[a]	13 900–28 000	3 000–10 000[b]
Total solids (ppm)[a]	13 150	2 900[b]
Biological oxygen demand, BOD (ppm) 5 days at 20°C[c]	1 800–2 920	1 250–2 200
Total solids (ppm)[c]	4 960	4 260
Biological oxygen demand, BOD (ppm) 3 days at 26·7°C[d]	2 400	3 900
Approximate volume of water per tonne of clean coffee (litres)		2 000–45 000[a] 8 000–21 000[c] 60 000–88 000[d]

[a] Aguirre et al.[23]
[b] Includes washing waters.
[c] Ward.[24]
[d] Brandon.[25]

skill in its operation, and generates a considerable waste disposal problem in the form of pulp and wastewater.

An approximate mass balance of a wet processing operation is presented in Fig. 1. Analytical details of the solid wastes are presented in Table 1 and for the wastewater in Table 2.

1.2.1. Wastewater Oxidation

Waters used in the fluming and pulping of cherries and in the fermentation and washing stages all contribute to the total volume of wastewater produced. This can be extremely variable (see Table 2), but with current processing techniques and some recirculation, water consumption can be reduced from its traditional value of around 20 000 gal (91 000 litres) per tonne of clean coffee produced to 3000 gal (13 650 litres).[26]

The pollution potential of the processing wastewater is not the result of any inherent toxicity but is due to microbial degradation of solutes and suspended particles, which reduces the level of dissolved oxygen in any receiving watercourse. It is this demand for oxygen exerted by a pollutant that forms the basis of one of the standard measures of pollution—the biological oxygen demand (BOD).[27]

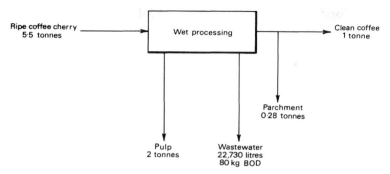

Fig. 1. Model coffee processing factory—mass balance.

The aerobic growth of micro-organisms on an organic material can be represented by:

$$CH_aO_b + O_2 + \text{minerals} + \text{biomass} \rightarrow CO_2 + H_2O + \text{dry biomass} + \text{heat}^{28}$$

In the case of a simple sugar such as glucose, a component of the ripe coffee cherry, the carbon balance becomes:

$$C_6H_{12}O_6 + O_2 \rightarrow CO_2 + H_2O + \text{dry biomass} + \text{heat}$$
$$\text{kg} \quad 1\cdot00 \quad 0\cdot53 \quad 0\cdot72 \quad 0\cdot60 \quad 0\cdot40 \quad 1520\,\text{kcal}$$

Thus, a large proportion of the organic material is oxidised to carbon dioxide, most of which will be lost in the gaseous form, and the remainder is converted to microbial biomass which will act as a source of nutrients to higher organisms.

This process occurs naturally when wastewater is discharged into a watercourse, provided its capacity for self-purification is not exceeded. If, however, the concentration of organic matter is too high, oxygen is removed from solution faster than it can be replaced by natural dissolution. In this event the level of dissolved oxygen will fall to zero, the river will become anaerobic and foul smelling, and its resource value will be severely reduced.

This is frequently observed as a result of the disposal of coffee wastewaters into rivers. A survey of several rivers between Nairobi and Thika in Kenya during the 1973–74 season showed that they were all seriously polluted with coffee wastes. Every river surveyed was anaerobic, and BOD values greater than 100 ppm were recorded.[26] An unpolluted river would normally have a BOD of less than 3 ppm, and one with a value of 12 ppm would be classed as grossly polluted.[29]

The pollution potential of an effluent is sometimes expressed more vividly in terms of an equivalent quantity of domestic sewage or the population from which this would arise. Thus the pollution resulting from the production of 1 tonne of clean coffee has been variously estimated as being equivalent to 273 000 litres (60 000 gal) of crude domestic sewage or to the daily sewage output of 2000 people.[25,26] This device should be viewed with some caution, however, since sewage is far more important than coffee wastewater as a potential source of human pathogens, and its volume and strength can vary considerably.[30]

Wastewater oxidation systems mimic the aerobic self-purification of natural watercourses but are engineered to supply oxygen at a rate appropriate to the wastewater's oxygen demand. Two main types of system can be distinguished: those in which the micro-organisms are adsorbed to some form of support material as a film, and those in which the microflora is in suspension.

The biological filter, also known as the percolating or trickling filter, was the first type of fixed film system introduced at the end of the 19th century for the treatment of sewage.[31] In the parlance of biotechnology these are classified as fixed-bed microbial film reactors. A settled liquid effluent is distributed uniformly over the surface of a bed of some inert material and is allowed to trickle through against a countercurrent of air. A microbial film develops, adsorbed on to the support medium, and purifies the effluent by a combination of mechanical entrapment, flocculation and biological oxidation.[32] The effluent from the filter is then subjected to a secondary sedimentation and either discharged or recirculated.

As with natural bio-oxidation, the process involves a complex ecosystem of organisms operating at several trophic levels and includes heterotrophic bacteria and fungi, holozoic protozoa, rotifers, nematodes and insect larvae and worms. The grazing activity of the latter, known as scouring organisms,[32] plays an important role in controlling the growth of the film which develops at a rate of 0.3–0.5 kg kg^{-1} BOD removed and would otherwise quickly block the filter.

Traditionally, mineral support materials such as rocks or gravel have been used, graded to contain pieces of approximately constant size so that the void space is maximised. More recently, plastic support materials have been introduced that have greater resistance to weathering and a higher surface area to volume ratio and voidage than mineral supports. This improves performance by giving better oxygen transfer and a larger area of microbe/effluent contact, thus permitting the use of a smaller filter for a given requirement. They are light in weight and can be formed into relatively tall towers economising on space but without expensive support structures.

The use of biological filters to treat coffee processing wastewaters was first reported 40 years ago but has not been widely adopted.[25,33] Brandon demonstrated that using a filter containing 38 m^3 of mineral packing and a loading rate of approximately 0·1 kg BOD m^{-3} per day, a reduction in BOD of greater than 94% was possible. The wastewater loaded on to the filter was diluted with 4 volumes of recirculated filter effluent to adjust the BOD to around 300 ppm.[25] With a plastic medium, manufacturers' information claims that a 90–95% reduction in BOD of a 300 ppm BOD effluent can be achieved using a loading of 0·25–0·4 kg BOD m^{-3} per day. Higher rates of BOD loading can be used, particularly with plastic media, in what is known as high-rate treatment. Generally the loadings used are greater than 0·6 kg BOD m^{-3} per day, often ten times those used in the conventional process, although the BOD reduction achieved is lower, normally in the range 50–80%.[34]

Biological filters are relatively simple to design and operate, have low running costs, a long life (30–40 years) and are very stable under continuous operation.[31] However, they do require a larger land area than some other procedures and sometimes give rise to a fly nuisance. Nevertheless, the success of such systems can be gauged from the estimate that they are employed in 70% of European and American effluent treatment plants.[31] A traditional biological filter to process the effluent from a factory producing 1 tonne of clean coffee per day would occupy about 400 m^2 of land and contain 500–800 m^3 of packing material.[35] By using a two-stage filter system of higher aspect ratio containing 220 m^3 of plastic packing material, this requirement could be substantially reduced. Both systems would, however, require a settling tank to remove sludge solids, 'humus', from the filter effluent.

A more recent development of the fixed film procedure is the rotating biological contactor (RBC). The microbial film is attached to the surface of plastic vanes, 1–3 m in diameter, rotating at a rate of 1–3 rpm and partly immersed in a trough through which the settled wastewater is passed. The natural buoyancy of the vanes partially offsets the weight of the unit so, although an electricity supply is necessary, the power requirement is low. One particular advantage of the RBC cited by Eden[32] is the fact that the periodic immersion of the film prevents it from being colonised by flies and their larvae thus avoiding the occasional fly nuisance associated with biological filters. Some calculations have indicated that for a 95% BOD reduction in a 300 ppm effluent the loading rate should be around $6\,g\,BOD\,m^{-2}$ per day.[36] Manufacturers, however, claim that loading rates of $12-16\,g\,BOD\,m^{-2}$ per day can be used. Loadings higher than around $20\,g\,BOD\,m^{-2}$ per day produce partial purification comparable to high-rate filtration.[34]

The system was originally developed for the treatment of sewage from small communities but has been increasingly applied to the processing of industrial wastewaters such as those from dairying.[37] An RBC is reported to be more resistant to shock loadings than a biological filter and to be more compact. A unit to treat the effluent from 1 tonne of clean coffee would occupy, not including settling tanks, $40-120\,m^2$ depending on the loading rate used and the quality of the final effluent required. They do have the disadvantage that they must be purchased as complete factory-built units.

In the other principal form of wastewater oxidation, generally known as activated sludge processes, micro-organisms grow as flocs dispersed in an aerated wastewater. These flocs purify the wastewater through a combination of adsorption and bio-oxidation, new cell material being generated at a rate of around $0.4\,kg\,kg^{-1}$ BOD removed. Processes normally operate as continuous culture systems with biomass feedback to maintain a sludge solids concentration in the range $1.5-8\,kg\,m^{-3}$ (see Fig. 2). The BOD loading of a conventional plant is around $0.5\,kg\,BOD\,kg^{-1}$ sludge solids per day.

Oxygenation of the wastewater which accounts for the considerable power consumption of these processes is achieved in a variety of ways. Direct diffusion of air bubbles introduced through porous diffusers submerged in the wastewater and mechanical entrainment are the most

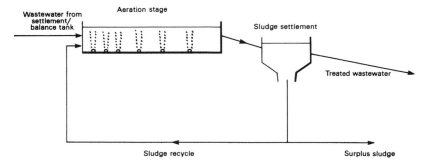

Fig. 2. The activated sludge process.

common. The latter, for instance, is used in the Pasveer or oxidation ditch in which effluent is aerated and driven around a closed-loop channel by a mechanical rotor.[38] The process consumes $1 \, kWh \, kg^{-1}$ BOD removed. The procedure, known as tapered aeration, attempts to reduce the high running costs of activated sludge processes by matching the aeration of the wastewater more closely to its oxygen demand. This is achieved by installing more diffusers in the tank near the inlet where the oxygen demand is greatest. A similar result can be achieved for mechanical aeration systems by introducing waste at intervals along the length of the tank.[31]

Oxygenation can be improved by increasing the partial pressure of oxygen in the gas phase, which increases the saturation concentration of dissolved oxygen. In the ICI deep shaft process this is achieved by introducing air into wastewater circulating in a shaft up to 150 m deep. The increased hydrostatic pressure as the wastewater passes down the central shaft or downcomer causes increased dissolution of the air. As the liquid passes up the concentric riser shaft, gases are released from solution causing a density change that maintains the liquid circulation. This technique enables a high concentration of active sludge to be maintained, facilitating high loading rates, typically $0.9 \, kg \, BOD \, kg^{-1}$ solids per day with a retention time of less than $2 \, h$.[34]

To our knowledge activated sludge plants have not been applied to the treatment of coffee wastewaters, the key factors probably being the high running costs and their susceptibility to shock loadings. They are more economical of space, however, and are said to have a lower capital cost than biological filters. One variation of the activated sludge process that may find some application in coffee processing is the use of surface aerators in an attempt to maintain the oxygenation of rivers receiving coffee wastewaters.

A major drawback of all wastewater oxidation systems is that optimum performance is obtained when they are operating continuously with uniform loadings. The seasonality of coffee processing would mean that an effluent treatment plant would have to be restarted at the beginning of each season and that a delay would be experienced in obtaining the required BOD reduction while the complex natural microflora developed. In addition, all the processes discussed here involve considerable financial outlay. They are applied to industrial food processing wastes in many countries, but only where punitive legislation controls the strength of discharges to watercourses and where disposal via the sewerage system incurs high treatment charges. A similar incentive would be required to induce the wholesale adoption of effluent treatment in coffee processing, and this would inevitably result in higher production costs.

1.2.2. Stabilisation Ponds

Stabilisation ponds are simple and effective means of effluent treatment most suited to situations where land is inexpensive, where organic loadings may fluctuate and where there is a shortage of technically trained personnel.[39] In such cases, Gloyna estimates that the costs involved are usually less than half those of other methods since ponds can be readily constructed without the importation of expensive equipment and have minimal running costs.[39] Lagoons have been shown to have considerable economic advantages in the treatment of potato-processing wastewaters in North America.[40] Their use in the treatment of coffee processing wastewaters has been discussed by several authorities,[24,41,42] and where sufficient land is available they appear to be one of the treatment options most readily applied.

The type of pond system required will depend upon the reason for treating the wastewater. In the case of coffee processing where a reduction of the BOD is the prime consideration, some combination of anaerobic and facultative ponds is necessary.

The anaerobic pond is used for the pre-treatment of high BOD wastes which would otherwise overload facultative ponds. In essence an open septic tank, it has a low surface area to volume ratio, a high BOD loading (greater than $0 \cdot 1 \text{ kg m}^{-3}$ per day), and, in the Tropics, a liquid retention time of 1–5 days.[38,39] Reduction in BOD occurs partly through sedimentation but largely as a result of degradation of organic carbon by facultative and anaerobic bacteria to produce a mixture of methane and carbon dioxide (see section 1.2.11).

Two disadvantages of the anaerobic pond are the frequent desludging required and the fact that it is often the source of unpleasant odours. The latter commonly occur when the pond is overloaded causing a drop in pH and an increase in the release of hydrogen sulphide and volatile acids. With careful control of the loading rate an alkaline pH of around 8 can be established and the problem of odours minimised.

Using the design equation of Vincent cited by Gloyna,[39] a 50% reduction in BOD can be expected from an anaerobic pond operating at 22°C and with a retention time of 5 days. Some caution should be exercised in applying these data to coffee wastewaters since they apply primarily to sewage in which the contribution of settlement to BOD reduction may be greater. However, assuming a depth of 2 m, an anaerobic pond to treat the effluent from the daily production of 1 tonne of clean coffee would occupy an area of approximately 57 m². The effluent from this pond would still have an extremely high BOD and would be a serious source of pollution if discharged into a watercourse without further treatment.

Facultative ponds rely on the concerted action of three classes of microorganisms. Anaerobic bacteria operate in the sediment at the bottom, degrading organic material to methane and carbon dioxide. In the middle zone of the pond, organotrophic organisms purify the effluent oxidatively, the oxygen supply being maintained by natural surface dissolution and by the activity of algae. This dependence on the activity of photosynthetic organisms results in diurnal variations in the conditions in the pond, dissolved oxygen and pH both decreasing during the hours of darkness. Continuous oxygenation to produce an activated sludge type of system can be achieved by the use of mechanical surface aerators.

The normal depth of a facultative pond in tropical conditions is around 1 m. If the pond is much deeper than this, it becomes largely anaerobic whereas in shallower ponds vegetation can grow up from the pond bottom creating conditions favourable for mosquito breeding.

The loading of facultative ponds is normally described in terms of the BOD applied per unit area; a rate of 112 kg ha^{-1} being recommended for ponds operating in tropical conditions.[43] Thus, a pond to treat the 80 kg of BOD resulting from the production of a tonne of coffee would occupy 0·7 ha (7000 m²). This requirement could be reduced considerably by using a pre-treatment anaerobic pond and facultative pond in series. In Australia the ratio between the area of facultative and anaerobic ponds is usually between 5:1 and 10:1, achieving overall BOD reductions of 65–80% during summer months.[39]

Seepage pits, which have been recommended for the disposal of coffee wastewaters in Kenya, combine a ponding treatment with land disposal and provide an effective and economical means of treatment so long as they are not overloaded.

1.2.3. Land Disposal of Wastewaters

Land has a natural capacity for effluent purification which, correctly managed, can provide the cheapest effective method for treating seasonal wastes (see, for example, references 43–45).

A variety of mechanisms contribute to the reduction in BOD that takes place as an effluent percolates through soil, including precipitation, adsorption and microbial degradation. The last is usually regarded as the most important factor.[43] The top layer of soil where conditions are aerobic acts as a natural biological filter converting organic material into biomass and minerals while at deeper levels anaerobic deomposition takes place.

The wastewater may be distributed over the land by a system of shallow furrows, or an area may be periodically flooded with wastewater which is then allowed to seep away. A spray system such as sprinkler heads distributed along a length of portable pipe or a rotating spray is more expensive but requires less maintenance than a furrow system.[46]

Land irrigation is used for the disposal of coffee processing wastewaters, and its extension offers a simple remedy for wastewater pollution while returning water and essential nutrients to the soil. Coffee wastewaters have been shown to be suitable for the irrigation of coffee trees and Napier grass, although the crop irrigated is of less importance than the physical characteristics of the receiving soil. Application of a liquid waste at rates higher than the capacity of the soil to deal with them will result in waterlogging, anaerobiosis, odour production and inhibition of plant growth. It may also lead to the run-off of highly polluting wastes into nearby watercourses. Ward noted some of these problems when a high-solids, ponded effluent was used for irrigation in El Salvador, possibly due to crusting over and masking of the soil, but a well sedimented wastewater showed no evidence of toxicity to plant life.[24]

Ideally the land used should slope gently to prevent ponding but not so steeply as to cause problems with erosion. Although it is difficult to generalise, roughly 1 ha of land is required for the disposal of a flow of $25\,m^3\,day^{-1}$.[46]

1.2.4. Wastewater Precipitation

In Kenya, Brandon tested a variety of chemical coagulants for their ability to reduce the BOD of mixed coffee wastewaters.[25] The results (Table 3) showed that calcium hydroxide, ferric chloride and aluminium sulphate all produced a reduction in BOD of around 48–49%. When the pollution potential was measured in terms of oxygen absorbed from permanganate, ferric chloride gave the best performance with a reduction of 18%. These rather modest improvements suggest that chemical treatment could make a contribution to coffee wastewater treatment but only in conjunction with some subsequent treatment.

An additional problem noted was that the sludge produced on treatment was frequently bulky and would not settle rapidly. This could probably be overcome nowadays by the use of polyelectrolyte flocculating agents.[47]

Table 3
The effect of coagulants on coffee wastewaters, after Brandon[25]

Coagulant (ppm)	Volume of settled sludge	pH after treatment	BOD reduction (%)
Calcium hydroxide			
400	30	9·6	40
800	15	10·0	49
Ferrous sulphate			
400	18	6·0	30
800	27	5·8	33
Ferric chloride			
200	15	4·8	40
400	25	3·9	49
800	15	3·5	40
Aluminium sulphate			
200	11	5·1	17
400	40	4·7	2
800	35	4·5	48

Orozco, working in Costa Rica, removed the precipitate obtained with chemical treatment by a rapid filtration step. Reasoning that a large proportion of the pollution potential of coffee wastewater was mucilage pectin, he precipitated it from the wastewater as calcium pectate by the stepwise addition of calcium oxide to a final level of 1% w/v. By this procedure he obtained reductions in the chemical oxygen demand (COD) of 80–90% and a final effluent COD of 420 ppm.[48] Using this procedure the time elapsed prior to treatment is likely to be of considerable importance since delay will result in increased degradation of pectin to compounds not precipitated by calcium ions.

1.2.5. Composting of Pulp

Once its natural physical barriers to infection are broken during pulping, microbial decomposition of coffee pulp quickly ensues. This occurs spontaneously, and uncontrolled in large quantities of pulp can lead to severe problems through the proliferation of flies, foul odours and polluting run-off. The fermentation of pulp spread in the field may also damage crops through acid formation and local heat generation.[4]

Composting is a system of solid waste management in which environmental conditions are optimised to accelerate the process of decomposition. Correctly operated it produces quickly and without nuisance a bland final product that can be easily handled, stored and applied to land without adverse effects.[49]

The composting of coffee pulp has been described by several authors.[50-53] The product should be odourless and less bulky than fresh pulp while retaining much of its nitrogen content. Since in several of the published descriptions no measures are taken to maintain a supply of oxygen to micro-organisms operating on the interior of the heap, the process would be essentially a natural, largely anaerobic, rotting of the pulp. A properly managed composting operation would achieve faster rates without the generation of an odour or fly nuisance by ensuring that aerobic micro-organisms can operate throughout the heap of material.

Several mechanical systems for aerobic composting have been described but the simplest procedure is that of windrow composting.[49] Material is piled into a long mound with a trapezoidal cross-section, 1 m high by 1 m across at the top and 1·5 m across at the base. Aeration is maintained by periodic turning of the heap which replaces exhausted air trapped in the interstices. The labour cost involved in such an operation should not prove excessive as it has been estimated that 4–6 tonnes can be turned in one man-day.[49]

Two intrinsic properties of the material to be composted, particle size and moisture content, have an important influence on aeration by determining the size and number of pockets in which air can be trapped. The most suitable mean particle dimension has been given as between 1 and 5 cm. Coffee pulp is at the lower end of this range, below which forced aeration has been recommended.[54] The high moisture content of coffee pulp (greater than 70%) means that heaps of pulp become sodden and closely packed producing an acid, anaerobic fermentation.[50] This can be overcome by a high frequency of turning, at least once a day during the early stages, or by the inclusion of some other material such as straw or coffee parchment to provide structural support and lower the overall moisture content. Golueke recommends turning every day until the moisture content is below 70%, at 2-day intervals when the moisture is 60–70% and at 3-day intervals if it is between 40 and 60%. If the moisture content is below 40%, it is necessary to add water.[55] This may be the case in the later stages of the fermentation when evaporative losses exceed the rate of water production by microbial metabolism.[56]

The balance between carbon and nitrogen in a substrate is one factor that determines the speed of composting, which can take from 2 to 10 weeks. A typical microbial cell contains 50% carbon and 5–12% nitrogen, and will fix only about one-third of the available carbon in a substrate as cell material, the other two-thirds being lost as CO_2. This suggests an optimal substrate C/N ratio in the range 10:1 to 30:1. In practice, where different carbon and nitrogen sources are not equally readily assimilated, the optimum is found to be 25–30:1.[49] If the ratio exceeds this value, then composting is slower since it takes several cycles of microbial growth to reduce it to a suitable level. With C/N ratios below 20:1, some losses in nitrogen may occur in the early stages of the fermentation when the pH can be around 8 to 9 and the considerable heat generated by microbial metabolism and trapped in the heap causes the temperature to rise above 60°C. Our own analyses of samples of pulp from Guatemala and Kenya have shown that C/N ratios, to be suitable for composting, vary between 22·4 and 27·9.

Compost should be used as a soil conditioner rather than as a fertiliser;[43,49] coffee pulp compost itself contains about 3·5% nitrogen.[6] Applied in conjunction with commercial fertilisers compost can prevent nutrient leaching by complexing inorganic phosphate and converting nitrogen into bacterial protoplasm which is slowly released as the bacteria die and decompose.[55] More importantly, compost has the physical effect of improving the air–water relationship of the soil by increasing its water retention capacity.[55] Beneficial effects resulting from the application of coffee pulp compost to crops have been noted,[4,57] as has the occasional absence of any observable effect.[58] Nevertheless, the use of coffee pulp compost should be regarded as a wise measure to conserve, if not improve, the long-term quality of the soil.

1.2.6. Production of Microbial Biomass

To the optimistically inclined, agricultural wastes represent a cheap and renewable raw material for conversion into useful products. One particular product that has attracted considerable attention is microbial biomass or single-cell protein (SCP) (see, for example, references 59–61).

The possibility of using coffee wastes as a substrate for SCP has not been neglected. One of the earliest of these investigations was by the Scientific Committee on Food Policy of the UK's Royal Society, which explored the question of food yeast manufacture. Coffee pulp was one of several raw materials which were eventually abandoned after consideration because of their localised and seasonal availability.[62]

The production of the yeast *Candida utilis* on a semi-continuous basis from coffee pulp and mucilage has been described.[3,51] The fermentation was carried out in 200-litre oak barrels which were unstirred but aerated at a rate of 500 litres min^{-1}. Wooden vessels are still widely used in fermentation industries producing vinegar and alcohol, but are difficult to clean and sterilise effectively. This, and the susceptibility of yeast propagation to infection, would argue strongly against their use in this case, although no problems with contamination were recorded.

The production medium was prepared by extracting the pulp obtained by dry pulping 300 kg of ripe cherry with water for 30 min. The mucilage-covered beans were treated with a pectolytic enzyme and then washed with water. The combined aqueous extracts contained 3% sugar and had a pH of 4·5–5·0.[3,51] The medium was then sterilised by the injection of steam, and mineral nutrient supplements were added (ammonium sulphate, superphosphate and potassium chloride). When the medium had cooled to 35°C, the yeast inoculum, prepared in two stages from a pure culture, was added. The course of the fermentation was followed by refractometry and microscopic examination of the broth. When the yeast concentration had reached its maximum, one-third of the vessel contents was removed and replaced with fresh medium.[3,51]

The average yield of dry yeast during 3 months' operation was 750 g for every 100 kg of whole coffee cherry processed. The crude protein content of the yeast was 50·4% on a dry weight basis, comparable to yeast produced from other sources.[3] Production of the same organisms by a continuous process using a medium extracted from pulp or hulls is the subject of a United States patent.[63]

More efficient procedures for extracting a fermentable juice from coffee pulp have been developed.[64] Using a batch hydraulic press and pre-treating the pulp with live steam and a commercial pectolytic enzyme, juice yields close to 60% of the pulp fresh weight are attainable. This juice contains 75% of the hot water extractable sugars and 35% of the total pulp solids. A commercial screw press is also available (E. H. Bentall & Co. Ltd, England), which, without pre-treatment, gives a juice yield of 27%. This could doubtless be improved by pre-treatment since the purpose of pressing in this case was not to prepare a high-extract juice but to dewater in order to facilitate sun-drying of the pulp (see p. 282).

Some preliminary investigations of the suitability of pressed coffee pulp juice as a substrate for mould biomass production have been published.[65] In shake flasks the growth of the four different moulds tested was substantially improved by supplementation of pasteurised juice with ammonium nitrogen and inorganic phosphate. The optimal initial C/N ratio was in the range 8–14 and biomass levels of 50 g litre^{-1} could be obtained in 24 h.

The growth of moulds on coffee-processing wastewaters has been investigated with a view to reducing the effluent's BOD and recovering some of the costs through the sale of the biomass as animal feed.[23,66-68] A pilot plant built at a large *beneficio* in El Salvador consisted of one 23 m^3 sedimentation tank, one 75 m^3 equalising tank, and two cylindrical 19 m^3 fermentation tanks. All tanks were open to the atmosphere and made of brick and cement with an interior coating of epoxy-paint. The wastewaters were supplemented with ammonium sulphate and phosphoric acid and the initial pH was adjusted to 3·5 with sulphuric acid to exert some selectivity for mould growth.

Experiments with non-aseptic batch growth of what was later identified as a strain of *Trichoderma harzianum* (Rifai) showed a 62% reduction in the COD of the wastewater in 16 h and 70% after 24 h, the net increase in biomass being 3·0 and 3·4 g litre^{-1} respectively.[68] Not surprisingly, in view of the generally lower growth rates of moulds and the non-aseptic character of the fermentation, attempts to operate the process continuously failed due to the mould being quickly outgrown by bacterial and yeast contaminants.

SCP produced from coffee wastes is likely to be used only as an animal feed. Although they were at one time seen as an answer to world food problems, SCP products currently play a negligible role in the human diet. In addition to the problem of the cells' high nucleic acid content, there is the major obstacle of consumer acceptability to overcome. Schemes launched in Trinidad, Jamaica, Costa Rica and the Philippines to produce food yeast have all failed.[69]

As an animal feed ingredient SCP would be subject to the least cost accounting of the feed compounder and would have to compete with other protein feedstuffs such as soya and fishmeal (see, for example, reference 70). In order to approach the price of these commodities it has been necessary for those commercial SCP plants that have been designed and built to take advantage of the economies of scale that arise from the use of very large plants producing of the order of 10^5 tonnes per annum. To append a small SCP operation on to a coffee processing plant is unlikely to be economic. On the basis of the published data a factory producing 1 tonne of clean coffee per day would produce sufficient solid wastes to produce 41·3 kg of yeast dry matter per day. In addition, there would be high running costs in the form of skilled personnel to operate the process and power for aeration, drying and ancillary operations, plus a special marketing effort to sell the product.

One variation of this approach that shows some potential is that of moist-solids fermentation of the coffee pulp to improve its nutritional characteristics (see section 1.2.12).[71]

1.2.7. Ethanol and Wine from Pulp

Ethanol is a chemical with a wide variety of uses. It functions as the physiologically active ingredient in alcoholic beverages, as a solvent or reactant in chemical processes and products, and as a liquid fuel. The fermentation of carbohydrate-containing materials by yeasts has traditionally served as the principal source of ethanol though the petrochemical-derived product is nowadays of great importance in non-potable uses. Around 7–8 million tonnes of ethanol are produced annually, approximately two-thirds by fermentation.[72]

The practical yield of ethanol from carbohydrates such as sucrose, glucose and fructose, the principal fermentable sugars in coffee pulp, is usually of the order of 516–582 litres per tonne dry matter. A factory producing 1 tonne of clean coffee per day would produce 2 tonnes of pulp. Assuming a pulp moisture content of 76·7% and that the total sugar content of 14·4% is fermentable,[11] this would yield 34–39 litres of anhydrous ethanol per day. A process described by Calle which includes an alkali treatment of the mucilage is recorded as giving a substantially higher yield of 55 litres per tonne clean coffee.[3]

Despite some research interest, the production of ethanol from coffee pulp has not been adopted on a practical scale since the process is unattractive from an economic standpoint. Below concentrations of about 6% w/v the amount of energy required to recover ethanol by distillation increases steeply.[73] Fermented pulp juice contains only 2·5–3·0% w/v ethanol so the recovery costs would be high.[74] This would be particularly important if the product was intended to compete in petrochemical ethanol markets, but less so if a potable spirit is produced. In the latter case, however, unpredictable factors such as consumer acceptability would be an important consideration.

The production of ethanol by fermentation also produces 12–15 volumes of its own wastewater with BOD values around 20 000 ppm for every volume of alcohol produced.[72] This would add to the pollution problems associated with wet processing. It also involves the use of plant with a relatively high capital cost.[72] Because of the seasonality of coffee processing, this would be underutilised unless alternative substrates are available for use outside the coffee-processing season.

Wine making is possible with a relatively modest capital outlay, and a dry white wine has been produced from coffee pulp on an experimental basis in Central America.[75-77] The product has not, however, been developed commercially. Organoleptic qualities apart, one problem is likely to have been the relatively high alcohol content of wines compared with fermented pulp juice. To achieve the minimum of around 8–9% w/v ethanol found in table wine it would have been necessary to add substantial quantities of sugar to the juice.

Ripe coffee berries are used in the Yemen and Somalia to produce a non-fermented drink known as kisher (*ghishr*) by boiling for several hours with a small amount of sugar and cardamom. It is possible that a similar product could be developed using separated pulp, but its market is likely to be small.

1.2.8. Pulp Vinegar

Vinegar is the product of a two-stage fermentation process. The first stage is a yeast fermentation that converts sugar into ethanol, and in the second, acetification, the ethanol is oxidised by bacteria to acetic acid.

In many countries there is a minimum legal standard for the concentration of acetic acid in vinegar of 4% w/v, though sometimes it can be higher. To achieve this minimum level it is necessary to start with a solution containing at least 8% w/v fermentable sugar and to have an intermediate alcoholic solution containing 4% w/v ethanol.[78] As already noted, this may be a problem with coffee pulp.

Nevertheless a traditional small-scale industry producing vinegar in Brazil has been described.[79,80] The larger factories produced 150–250 litres per day during the season using wet processed pulp. The acetification process employed was that of a trickling tower. The acetic acid bacteria which convert the ethanol in the fermented pulp juice into acetic acid were adsorbed on to an inert support material, coffee twigs, and the acetifying solution trickled down over the support against a countercurrent of air. The overall process yield was about 74% and no difficulty was reported in obtaining a satisfactory final acetic acid concentration (4·6% w/v). This may be due to the remarkably low moisture content of the pulp used (42·7%), or to a significant contribution from the alcohol employed in the initial sugar extraction step.

1.2.9. Pectin

Pectin is a polysaccharide consisting primarily of 1-4-linked D-galacturonic acid residues partially esterified with methanol. It is an important structural component of the tissues of higher plants and is exploited commercially for its gelling properties. Pectins with a high degree of esterification (60–75%) form an acid–sugar–pectin gel and are used in the production of jams and confectionery jellies. Products with a low degree of esterification (20–45%) form calcium pectinate gels, which are used in sugar-free or low-sugar jellies.[81] World production of pectin is estimated at around 10 000 tonnes per annum principally from citrus waste (75%) and apple pomace.[82] The possibility of using coffee pulp and mucilage as a source of pectin has received attention from several authors.

The most commonly quoted figures for the pectin content of coffee pulp and mucilage on a dry matter basis are 6·5% and 35·8% respectively;[3,83] yields of 8–12% based on the whole dried cherry have also been cited.[82] Rolz obtained a pectin yield of 17% from the mixed pulp/mucilage waste obtained when a screw press was substituted for the conventional coffee pulper.[22] The technology of pectin extraction is well established and should be readily applicable to coffee wastes. Calle has described the seven extraction procedures investigated at Cenicafe, Colombia, in the course of designing a pilot plant.[3] Trials conducted in the late 1960s at the Tropical Products Institute showed that pectin extracted from fresh pulp was generally of the 'slow set' type with a

Fig. 3. Official presentation of Kahawa coal during Vice-presidential visit.

degree of esterification in the range 59–62% (B. J. Francis, personal communication).

One problem with the use of pulp is the stabilisation of the pectin prior to its extraction. Unless microbial and enzymatic changes in the pulp can be arrested, considerable losses will occur through depolymerisation reactions which decrease the overall yield, and demethylation of the pectin which impairs its gelling properties. Drying is an effective means of achieving this, although sun-drying, being relatively slow, is still likely to result in product losses. Techniques such as alcohol treatment or blanching in boiling water may also be used.[82]

An analysis conducted in 1975 of the economics of establishing commercial pectin production in developing countries was not particularly sanguine, pointing out that there was no likelihood of any shortage in the sources currently used.[82] It did, however, note that this situation was dependent on factors which may be subject to change. The essential prerequisite of any proposed programme would therefore have to be a realistic projected cost analysis in addition to a product of assured quality and availability.

1.2.10. Parchment Charcoal

The parchment and silverskin removed in the last stage of wet coffee processing does not present a major disposal problem. In Kenya, farmers have traditionally collected parchment and spread it in cattle pens where it is trampled in to produce a manure that is used in the coffee plantations.

Surpluses can also be burnt as fuel and in some countries quantities are sold to heat industrial boilers. As noted previously, the transport costs involved can be considerably reduced by conversion of agricultural residues to charcoal (see section 1.1.1). In the 1970s the Kenya Planters' Cooperative Union (KPCU) developed the production of charcoal briquettes from parchment for domestic fuel use. The husks are carbonised in a kiln, ground, coagulated and then moulded into briquettes prior to packing into 2, 4 and 50 kg bags (Fig. 3). The product, 'Kahawa coal', has longer burning characteristics than conventionally produced wood charcoal, presumably due to a greater density, and is being heavily promoted in both urban and rural areas as a means of decreasing the alarming rate of deforestation.

1.2.11. Biogas Production

Anaerobic digestion of an organic material converts the fixed carbon into microbial biomass and a mixture of carbon dioxide and methane termed biogas. Since there is no need to oxygenate and the rate of biomass production is less than that observed in aerobic processes, anaerobic digestion represents an extremely effective way of reducing the BOD of high-strength effluents. Although reductions of up to 90% or more are possible, anaerobic digestion is usually employed as a pre-treatment measure because the effluent it produces is not normally suitable for surface-water discharge without subsequent treatment. It is applied in its simplest form in the anaerobic lagoons described earlier, and is widely used in the treatment of sewage to reduce and stabilise the sludges separated from the sewage and produced during aerobic treatment.

By carrying out the fermentation in a completely enclosed vessel, the increased capital cost can often be more than offset by the fuel value of the biogas collected. In the case of sewage, the evolved gas comprises approximately 30% CO_2 and 70% methane with an overall calorific value of 26 MJ m^{-3} making 1·3 m^3 of gas equivalent to 1 litre of petrol.[84] The major London sewage works, like others around the world, use the 60 000–65 000 m^3 day^{-1} of biogas they produce as a source of fuel to power their operation.[85]

Advocacy of the wider potential of biogas production as a source of renewable fuel, particularly in rural areas, is by no means a recent phenomenon although the oil price increases in the 1970s gave a new impetus to the field.[86] As a result numerous surveys and practical guides have been published over the last 10 years or so (see, for example, references 87–89). However, as van Brakel has pointed out, much of the work done in this area is semi-empirical; there has not been steady and uninterrupted progress towards agreed conclusions, principles, design or practice, and recent publications do not necessarily supersede or improve on earlier ones.[86]

The complete anaerobic digestion of a hypothetical organic compound can be represented as:

$$CH_aO_b + \text{minerals} + \text{biomass} \rightarrow$$
$$CO_2 + CH_4 \pm H_2O + \text{more biomass} + \text{heat}$$

In the case of glucose the carbon mass balance becomes:

$$C_6H_{12}O_6 \rightarrow CH_4 + CO_2 + \text{dry biomass} + \text{heat}$$
$$\text{kg} \quad 1\cdot00 \quad\quad 0\cdot25 \quad 0\cdot69 \quad\quad 0\cdot056 \quad\quad +89\cdot1\,\text{kcal}^{28}$$

The overall conversion is due to the concerted action of two groups of micro-organisms. The first group of obligate and facultatively anaerobic organisms degrade complex organic molecules into simple compounds such as acetate and carbon dioxide. These then serve as substrates for the methanogenic bacteria, slow-growing obligate anaerobes that convert them to methane and, in the case of acetate, carbon dioxide.

The slow growth rate of the methanogenic bacteria is a major limitation of the process. To avoid washing them out from conventional continuous digesters, residence times of 20–30 days are required. With large effluent flows this necessitates very large and costly vessels. Various procedures have been developed that accelerate the rate of digestion and reduce the retention time. In the contact process, for instance, biomass is separated in a settling tank and recycled to the main fermenter which can then be operated with a far shorter hydraulic retention time (HRT). Crucial to the successful operation of this technique is the ease with which the biomass can be separated from the fermenter effluent.[90]

Generally, anaerobic digesters have a reputation for instability and sudden failure, particularly in response to a change in loading rate. This appears to stem from the difficulty in establishing a satisfactory balance between the activity of degradative bacteria and methanogenesis. The former tend to decrease the pH of the medium through the production of volatile acids like acetate. The methanogenic bacteria are susceptible to low pH values and function poorly below pH 6·6 and not at all below 6·0. In a balanced fermentation the levels of pH and volatile acidity will both remain relatively constant at around 6·7–7·4 and 300–500 ppm (as acetate), respectively. A sudden increase in the volatile acidity is a characteristic sign of imminent digester failure unless remedial action is taken. This problem is reduced when using well buffered media such as cattle manure, but is particularly severe with materials such as coffee pulp, fruit and vegetable wastes which have little endogenous buffering capacity in addition to quantities of readily degraded carbohydrates. The difficulties encountered in the scale-up and commissioning of a digester to treat the effluent from starch processing have been described by Morgan,[91] and laboratory trials conducted with fruit and vegetable wastes containing similar levels of soluble sugar to fresh coffee pulp were found to require the addition of alkali to control the pH within an acceptable range.[92]

Coffee pulp, which has been the subject of both laboratory and practical trials,[3,93–97] presents similar problems. T. H. Hutchinson of Tunnel Estate, Fort Ternan, Kenya, has pioneered the use of methane digesters on farms in East Africa and his company has sold about 150 units in the region. His experience with a batch digester has been that coffee pulp is a good gas producer only when mixed with cattle manure to buffer the medium against pH changes arising from the rapid catabolism of soluble sugars. Similar observations have been reported from Mexico using a semi-continuous process where a mixture of 84% pulp with 16% manure permitted operation with a retention time of 20 days and a loading rate of 3·6 kg volatile solids m^{-3} day.[98] Where successful trials have been conducted using coffee pulp without manure, it is interesting to note that the pulp used was several months old by which time rapidly metabolisable sugars would have disappeared.[99]

Calzada et al.[100] have experimented with the promising concept of using a two-stage process in which the activities of the two groups of organisms are physically separated. The acidogenic stage was operated at a pH of 3·2–3·8 with an HRT of 0·5 days and a loading rate of 56 kg volatile solids m^{-3} day. The effluent from this stage was adjusted to pH 7–8 with alkali before feeding to the methanogenic fermenter running with a retention time of 8 days and a loading rate of 1·8 kg volatile solids m^{-3} day.

Certainly anaerobic digestion is an attractive concept for the treatment of coffee wastes; it would combine the virtue of assisting the disposal of both pulp and wastewaters with a useful yield of gaseous fuel, potentially around 66 m^3 per tonne of pulp. However, in view of the current unreliability of the process, and the skilled attention required, the high capital cost of such plants may not be justified at present.

1.2.12. Coffee Pulp as an Animal Feed

Coffee pulp contains about 10% crude protein on a dry weight basis with a protein quality similar to that of soya.[11] Its use as an animal feed ingredient has been limited by its seasonal availability (2–3 months a year), its high water content and consequent perishability, and by the presence of anti-nutritional factors in the pulp which restrict the quantities that can be used in a ration.

Extensive trials have established that there is generally an inverse relationship between the level of coffee pulp in a diet and weight gain, and that this is more marked above a certain level. In extreme cases, particularly with smaller animals such as chickens, mortality has been shown to increase with the level of coffee pulp fed. In one feeding trial reported by Bressani and co-workers all 10 chicks fed 50% pulp died within 6 weeks.[101]

With cattle the maximum level is about 20%.[102] The poor performance of animals fed higher levels appears to be due to a reduction in the voluntary feed intake and interference with normal metabolism by components of the pulp. Although the precise nature of the anti-physiological action of coffee pulp has yet to be determined, it is thought to arise from the combined effects of caffeine, with its stimulatory and diuretic action, and phenolic compounds such as chlorogenic acid, caffeic acid and tannins which decrease protein availability and inhibit digestive enzymes and microorganisms in the gut. The relatively high levels of potassium present may also play some part.[103]

With monogastric animals there is an additional limitation due to the high fibre content of the pulp (about 21%). Levels of up to 16% pulp can be used in pig rations without any detrimental effect on weight gain and feed conversion, but even a 10% inclusion rate has been shown to have a negative effect on chickens.[101,104] Nevertheless substantial benefits could result from the limited use of pulp as an animal feed ingredient. From weaning to market a pig needs 320 kg of feed, and substitution of coffee pulp for corn at the 16% level would release 52 kg of grain per head for alternative use.

Some improvement in the nutritive value of coffee pulp for monogastric animals has been reported to result from a moist-solids fermentation with the mould *Aspergillus niger*.[71] The product when included in a ration for growing chickens at 10% had a feed efficiency close to that of the standard ration and significantly better than unprocessed pulp fed at the same level. This may result from the loss in dry weight as respired carbon dioxide that takes place during fermentation giving an increased protein content in the product on a dry weight basis. The level of lignin, tannin and caffeine remained similar, signifying some overall loss of those components, but probably the most significant factor was the substantial decrease in indigestible components measured as crude fibre, cellulose and hemicellulose.

Extraction of anti-nutritional factors would detoxify pulp and improve its value, and may also lead to the production of a marketable by-product, caffeine. Aqueous extraction for 1 h at 96°C with 20 parts solvent to 1 part dried pulp reduced the pulp caffeine content by 81·6%, the total solids by 44·3% and the tannins by 39·9%.[105] In feeding trials the product still had a lower protein digestibility than the normal ration, although nitrogen retention was improved. The absence of any dramatic improvement in the product's feed value was attributed to possible changes in protein availability that may have occurred during pre-drying of the pulp.[106] If this is the case, then extraction of the fresh pulp may prove advantageous.

The perishability of fresh pulp may be overcome by drying or ensiling. Solar drying is the most economical method available although, dependent as it is on the weather, it is not the most reliable. From data presented by Molina[107] it seems that at 60–65% relative humidity an 8-day drying period would normally be sufficient to reduce the pulp moisture content from 80–85% to below 15%. At a loading rate of 22·7 kg pulp m^{-2}, the 2 tonnes of pulp produced from the daily production of 1 tonne of clean coffee would require a total drying area of 705 m^2. This requirement could of course be substantially reduced by reducing the initial moisture content of the pulp by pressing (see p. 272).

Artificial drying can be expensive in terms of its fuel energy requirement although this may be offset by the use of parchment as a fuel. In San Jose, Costa Rica, Subproductos de Cafe have been operating a pulp dehydration plant since 1981. The product is sold to feed mills for compounding into pig, poultry and dairy cattle feeds.

In the production of silage an otherwise perishable material is preserved as the result of a lactic acid bacterial fermentation. A sequence of microorganisms of the genera *Streptococcus*, *Leuconostoc*, *Lactobacillus* and *Pediococcus* produce a mixture of lactic and acetic acids (1–1·5% by weight expressed as lactic acid) which decreases the pH to below 4·2 and inhibits the growth of spoilage organisms, particularly *Clostridium* spp. Commercial starters are available but usually the fermentation is a spontaneous one, the development of the correct microflora being favoured by ensuring anaerobic conditions and an adequate supply of fermentable sugar.[108,109] Molasses, a relatively cheap source of the latter, is often used as an additive to promote lactic fermentation, the usual recommended levels being 0·9–1·8% by fresh weight.[108] One disadvantage in this is the need to add water to the molasses to reduce its viscosity and facilitate mixing.[109]

In a procedure to produce coffee pulp silage, molasses was added at a rate of 3–5% diluted with 50% water. Prior to the molasses addition the pulp had had its moisture content reduced to 65–70% by partial sun-drying or pressing.[110]

Although the silo used for ensiling can be some form of permanent structure, a pit silo was used in this work because it was cheap to construct, takes up little space and gives the maximum degree of protection. The pits were 6–8 m by 12–15 m and 1·5–2·0 m deep. The interior was covered by plastic sheeting to protect the silage from soil and a drain was opened in the bottom. Pulp could be piled in the pit to a height of almost 2 m above ground, provided it was covered with plastic sheeting and soil to protect it from rain and maintain anaerobic conditions.

Ensiled coffee pulp has been found to be superior to dried coffee pulp in terms of its nutritive quality, but the *in vitro* digestibility of the dry matter is less. The improvement noted in the acceptability of coffee pulp after ensiling is thought to be due to the loss of water-soluble caffeine and tannins in the draining liquids.[110]

2. SECONDARY PROCESSING: THE PRODUCTION OF INSTANT COFFEE

The growth of the instant coffee market in recent years due to the modern consumer's desire for a convenience product has been impressive. This market expansion has also led to the production of increasing quantities of spent coffee grounds as an unwanted by-product of instant coffee manufacture. The economic disposal of large quantities of waste is an important factor in reducing plant operating costs.

2.1. Coffee Grounds

In a typical instant coffee manufacturing process, roasted ground beans are treated with water at high temperatures and pressures to extract the soluble material. The insoluble residue (grounds) is blown to silos at a moisture content of 75–80%. This slurry is then dewatered to a cake of about 50% moisture content by screw presses. The screw press is capable of dewatering only the coarsest coffee grounds, and fine material remains in the press effluent at an approximate concentration of 2–7% by weight, which must be reduced to an acceptable level prior to discharge as effluent.

A process has been developed (Pennwalt Ltd, Camberley, UK) in which the press effluent is fed into a continuous centrifuge. The solids recovery is claimed to be between 85 and 95%. The process is illustrated in Fig. 4.

It is difficult to obtain precise information on the scale of the problem of spent coffee ground disposal because of the reluctance of manufacturers to provide information that may indicate production levels and process yields. A perspective on the quantities of spent grounds produced is provided by the fact that 1 tonne of green coffee releases 480 kg of spent grounds assuming a 20% loss on roasting and a 40% extraction efficiency. Sivetz and Desrosier note that the combined output of three soluble coffee

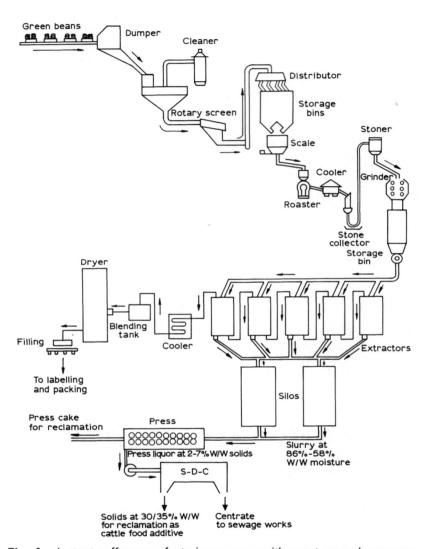

Fig. 4. Instant coffee manufacturing process with spent grounds recovery.

manufacturing plants in Brazil doubled over the 10 years up to 1977 to 33 600 kg of dry spent grounds daily.[111]

The most complete analysis of spent grounds has been provided by Blair[112] and is shown in Table 4. Spent coffee grounds are seen to be a fibrous high-energy by-product. The calorific value is comparable to that of charcoal. The protein level is of less significance since the heat treatment

Table 4
Composition of spent coffee grounds (after Blair[112])

Component	Per cent	Fatty acids (per cent of ether extractable lipid)		
Moisture	7–8	Palmitic acid	(16:0)	34·2
Crude protein	10–12	Palmitoleic acid	(16:1)	0·4
Crude fibre	35–44	Stearic acid	(18:0)	9·7
Lignin	Assessed at 36 of total fibre	Oleic acid	(18:1)	14·0
		Linoleic acid	(18:2)	37·0
		Arachidic acid	(20:0)	3·8
Cellulose	Assessed at 64 of total fibre	Gadoleic acid	(20:1)	0·4
		Behenic acid	(22:0)	0·2
		Free fatty acids		7·6
Ether extract	22–27	Iodine value		85–93
Nitrogen-free extractives	13–18	Saponification No.		185–193
	Mainly hemi-	Peroxide value		11–17
	cellulose and pentosans	Melting point (°C)		35
Ash	0·25–1·0	Amino acids (per cent of protein)		
Silica	0·2	Aspartic		4·1
Calcium	0·08	Threonine		2·8
Magnesium	0·01	Serine		1·9
Potassium	0·04	Glutamic acid		18·6
Sodium	0·03	Proline		7·2
Phosphorus	0·01	Glycine		7·6
Manganese	26·8 ppm	Alanine		6·2
Zinc	10·0 ppm	Valine		9·0
Copper	35·0 ppm	Methionine		2·0
Chlorine	Negative	Isoleucine		6·3
Selenium	0·26 ppm	Leucine		13·4
		Tyrosine		4·2
Miscellaneous		Phenylalanine		8·3
Tannic acid	0·9	Lysine		2·9
Gross energy	6930 kcal kg^{-1} dry matter basis	Histidine		2·2
		Arginine		trace
		Hydroxyproline		1·0
Starch equivalent	75 on ruminants			

that the bean undergoes during roasting results in a considerable reduction in the biological value of the protein. Furthermore, during the manufacture of instant coffee, the soluble protein is removed and only the insoluble portion remains. Spent grounds contain between 9 and 20% oil[112] and are rich in palmitic acid. The oil resembles groundnut oil in its characteristics but for the high unsaponifiable matter of 9–12%[113] which would pose a problem in the production of edible oil. The percentage of unsaturated fatty acids is greater than 50%; this high percentage corroborated by the iodine value of 85–108 would imply a significant risk of rancidity, although some protection would be afforded by the presence of natural antioxidants.

2.2. Spent Coffee Grounds as Fuel and Feed

A number of possible uses of spent grounds have been suggested which include sawdust substitute,[2] methanol and acetone[114] and more recently anaerobic digestion.[115] Industrial processing will only be viable if the residue is processed near the soluble coffee plant to avoid transportation of a large quantity of water. It is unlikely that any of the aforementioned uses have an economic value. There are, however, two solutions to the problem of disposal of spent grounds which are economically viable, viz. use as an animal feed or as a boiler fuel.

The idea of feeding animals with spent grounds is not a new one. Morgen in 1918 carried out trials incorporating 30% of spent coffee grounds in rations for ewes, but with little success.[116] It is only in recent years that the incorporation of spent grounds in animal rations has been fully investigated, and its use has expanded in conjunction with the development of instant coffee processing. In 1969 a commercial product 'Cherco' (Cherwell Valley Ltd, Banbury, Oxon, UK) was successfully introduced in the UK (Fig. 5). After pressing the spent grounds as previously described, the product is dried to a moisture content of 10–12% in a rotary drier. In view of the high fibre levels, dried grounds are normally recommended for use in ruminant feeds at levels up to 7·5%, and the slightly bitter taste limits inclusion levels to below 10% in complete feeds. Since 1970, following the rapid increase in available tonnage in Western Europe, approximately 40 000 tonnes per annum have been consumed in animal feeds, mainly in dairy and other ruminant rations. The year-round availability of the product is a considerable advantage to the animal feed compounder with regular deliveries resulting in low storage costs.

Recycling of spent coffee grounds as a fuel source is an alternative

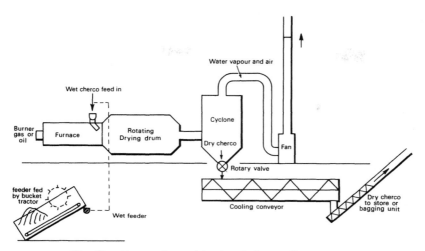

Fig. 5. Rotary drum drier for drying coffee residue.

attractive proposition in view of high energy costs at the time of writing. Following mechanical pressing of the waste grounds to a moisture content of 50%, the grounds are burnt as the base load in refractory cells with the addition of gas or oil to maintain the peak level of steam required for coffee processing. Coffee grounds are an ideal fuel in this environment-conscious age. Virtually smoke-free during combustion, the grounds burn to a low ash residue with low particulate emission in the flue gas. In view of the lightness of coffee ground ash, overfire rather than underfire is used in the combustion cells to minimise the loss of flyash particles into the airstream. Sivetz[116] has been critical of the design of most industrial grounds burning systems, and particularly of the failure to dewater the grounds adequately. The high calorific value of spent grounds reduces to 3300 kcal when the moisture content is 75%. The result has been that a great number of installations require fuel oil to burn up the grounds because the feed grounds are too wet. The grounds should be dried to around 33% for efficient fuel use, using any of several types of industrial dryer which are available. Drying to below 30% moisture cannot be done safely when high heating temperatures are used because of the fire risk.

REFERENCES

1. Martinez Nadal, N. G., *Coffee Tea Ind.*, 1958, **81** (3), 9, 10, 33, 34.
2. Tango, J. S., *Bol. Inst. Technol. Aliment. Sao Paulo*, 1971, **28**, 49–73.
3. Calle, H. V., *Bol. Tec. No. 6 Cenicafe, Colombia*, 1977, 84 pp.

4. Calude, B., *Cafe Cacao Thé*, 1979, **23**, 146–52.
5. Adams, M. R. and Dougan, J., *Proc. 9th Coll. ASIC*, 1980, 325–33.
6. Adams, M. R. and Dougan, J., *Trop. Sci.*, 1981, **23** (3), 177–96.
7. Rowe, J. F., *The World's Coffee*, HMSO, London, 1963.
8. Clarke, R. J., *Process Biochem.*, 1967, **2** (10), 15–19.
9. Clarke, R. J., *Tea Coffee Trade J.*, 1971, **140** (2), 14–16, 35, 36, 40, 41.
10. Singh, S., DeVries, J., Hulley, J. C. L. and Yeung, P., *World Bank Staff Occasional Papers* No. 22, Johns Hopkins University Press, Baltimore, 1977.
11. Bressani, R., Cabezas, M. T., Jarquin, R. and Murillo, B., in *Proc. Conf. Anim. Feeds Trop. Subtrop. Orig.*, Tropical Products Institute, London, 107–17.
12. Rolz, C., *J. Appl. Chem. Biotechnol.*, 1978, **28**, 321–39.
13. Sivetz, M. and Foote, H. E., *Coffee Processing Technology*, Avi, Westport, Conn., 1963.
14. Smith, A. E., *Trop Sci.*, 1985, **25**, 29–39.
15. Coombs, J., *Chem. Ind.*, 1981, 223–9.
16. Rogerson, A., *E. Afr. Agric. J.*, 1955, **20**, 254–5.
17. Ledger, H. P. and Tillman, A. D., *E. Afr. Agric. For. J.*, 1972, **36–37**, 234–6.
18. Ledger, H. P. and Tillman, A. D., *E. Afr. Agric. For. Res. Org. Ann. Rept*, 1971, 208.
19. Geevarghese, P. I., Subrahmanyam, M. and Ananthasubrahmaniam, C. R., *Kerala J. Vet. Sci.*, 1980, **12** (1), 11.
20. Geevarghese, P. I., Subrahmanyam, M. and Ananthasubrahmaniam, C. R., *Kerala J. Vet. Sci.*, 1981, **12** (2), 288–94.
21. Jones, K. L. and Jones, S. E., in *Progress in Industrial Microbiology*, Ed. M. E. Bushell, Elsevier, Amsterdam, 1984, 411–56.
22. Rolz, C., Menchu, J. F., Calzada, F., de Leon, R. and Garcia, R., *Process Biochem.*, 1982, **17** (3), 8–10, 22.
23. Aguirre, F., Maldonado, O., Rolz, C., Menchu, J. F., Espinosa, R. and de Cabrera, S., *Chem. Technol.*, 1976, **6**, 636–42.
24. Ward, P. C., *Sewage Wks J.*, 1945, **17**, 39–45.
25. Brandon, T. W., *E. Afr. Agric. J.*, 1949, **14**, 179–86.
26. Director of Kenya Water Dept, *Kenya Coffee*, 1977, **42**, 367–73.
27. Wheatstone, K. C., in *Treatment of Industrial Effluents*, Eds A. G. Callely, C. F. Forster and D. A. Stafford, Hodder & Stoughton, London, 1977, 30–64.
28. Pirt, S. J., *J. Appl. Chem. Biotechnol.*, 1978, **28**, 232–6.
29. Hinchcliffe, P. R., *Chem. Ind.*, 1980, **15**, 603–9.
30. Roberts, D. G. M., in *Water Pollution Control Technology*, HMSO, London, 1979, 22–36.
31. Wheatley, A. D., *Chem. Ind.*, 1982, **17**, 512–18.
32. Eden, G. E., in *Water Pollution Control Technology*, HMSO, London, 1979, 48–59.
33. Horton, R. K., Pacheco, M. and Santana, M. F., *Sewage Wks J.*, 1947, **19**, 534–8.
34. Winkler, M. A., *Biological Treatment of Waste-water*, Ellis Horwood, Chichester, 1981.

35. Mathew, P. K., *Indian Coffee*, 1978, **42**, 343–4.
36. Bruce, A. M. and Merkens, J. C., Water Research Centre, Stevenage, Laboratory Reprint No. 761, 1975.
37. Godfrey, R. E., *Fd Process. Ind.*, 1980, **49** (589), 31, 35, 37.
38. Mara, D. D., in *Water, Wastes and Health in Hot Climates*, Eds R. Feachem, McGarry and D. Mara, John Wiley, London, 1977.
39. Gloyna, E. F., *Waste Stabilization Ponds*, WHO, Geneva, 1971.
40. Cocci, A. A., McKim, M. P., Landine, R. C. and Viraraghavan, T., *Agric. Wastes*, 1980, **2**, 273–7.
41. Coffee Pollution Committee, *E. Afr. Agric. J.*, 1939, **4**, 370–7.
42. Pahren, H. A. and Saenz, R. F. Robert A. Taft Sanitary Engineering Centre Tech. Rep. W60-2, 1961.
43. Loehr, R. C., *Agricultural Waste Management*, Academic Press, New York, 1974.
44. Hobson, P. N. and Robertson, A. M., *Waste Treatment in Agriculture*, Applied Science Publishers, London, 1977.
45. Wood, B. J., Pillai, K. R. and Rajaratnam, J. A., *Agric. Wastes*, 1979, **1**, 103–27.
46. Mara, D. D., *Sewage Treatment in Hot Climates*, John Wiley, Chichester, 1978.
47. Hiebenthal, F. and Spetch, D., *Effluent and Waste Treatment J.*, 1967, **7** (6), 308–12.
48. Orozco, R. A., *Proc. 6th Coll. ASIC*, 1973, 290–6.
49. Golueke, C. G., *Biological Reclamation of Solid Wastes*, Rodale Press, Emmaus, PA, 1977.
50. Beckley, V. A., *Kenya Dept. Agric. Bull.*, 1934, No. 9, 45 pp.
51. Aguirre, F., *Invest. tecnol. del ICAITI*, 1966, No. 1.
52. Anon, *Indian Coffee*, 1974, **38**, 279–82.
53. Anon, *Indian Coffee*, 1976, **40**, 274.
54. Gray, K. R., Sherman, K. and Biddlestone, A. J., *Process Biochem.*, 1971, **6** (10), 22–8.
55. Golueke, C. G., in *Handbook of Solid Waste Management*, Ed. D. G. Wilson, Van Nostrand Reinhold, New York, 1977.
56. Finstein, M. S. and Morris, M. L., in *Advances in Applied Microbiology*, Vol. 19, Ed. D. Perlman, Academic Press, New York, 1975.
57. Parra, H., *Cenicafe*, 1959, **10**, 441–60.
58. Lopez Arana, M., *Cenicafe*, 1966, **17**, 121–31.
59. Laskin, A. I., in *Annual Reports on Fermentation Processes*, Vol. 1, Ed. D. Perlman, Academic Press, New York, 1977, 151–75.
60. Birch, G. G., Parker, K. J. and Worgan, J. T. (Eds), *Food from Waste*, Applied Science Publishers, London, 1976.
61. Rolz, C., in *Annual Reports on Fermentation Processes*, Vol. 7, Ed. G. Tsao, Academic Press, New York, 1984, 213–356.
62. Thaysen, A. C., in *Yeasts*, Ed. W. Roman, D. R. W. Junk, The Hague, 1957.
63. Fries, K. W. E., US Patent 3,576,720, 1971.
64. Rolz, C., Menchu, J. F., Del Carmen de Arriola, M. and de Micheo, F., *Agric. Wastes*, 1980, **2**, 207–14.

65. DeLeon, R., Calzada, F., Herrera, R. and Rolz, C., *J. Ferment. Technol.*, 1980, **58** (6), 579–82.
66. Rolz, C., Mayorga, H., Schneider, S., Cordero, L., Menchu, J. F., Espinosa, R. and Church, B., *Proc. 6th Coll. ASIC*, 1973, 260–1.
67. Rolz, C., in *Single-cell Protein*, Vol. 2, Eds S. R. Tannenbaum and D. I. C. Wang, MIT Press, Cambridge, Mass., 1975, 273–313.
68. Espinosa, R., Maldonado, O., Menchu, J. F. and Rolz, C., *Biotechnol. Bioengng Symp.* No. 7, 35–44, 1977.
69. Schrimshaw, N. S., in *Single-cell Protein*, Vol. 1, Eds R. I. Mateles and S. R. Tannenbaum, MIT Press, Cambridge, Mass., 1968, 3–7.
70. Thomka, S., in *Production and Feeding of Single-cell Protein*, Eds M. R. Ferranti and A. Fiechter, Applied Science Publishers, London, 1983.
71. Penaloza, W., Molina, M. R., Brenes, R. G. and Bressani, R., *Appl. Environ. Microbiol.*, 1985, **49** (2), 388–92.
72. Adams, M. R. and Flynn, G., *Rep. Trop. Prods Inst.* G169, v + 26 pp, 1982.
73. Parker, K. J. and Righelato, R. C., in *Processes for Chemicals from Some Renewable Raw Materials*, Institution of Chemical Engineers, London, 1979, 9–21.
74. Krishnamoorthy Bhat, P. and Deepak Singh, M. B., *J. Coffee Res.*, 1975, **5**, 71–2.
75. Anon, *Spice Mill*, 1934, **57**, 313.
76. Anon, *Revista agric. Guatemala*, 1940, **17**, 179–80.
77. Marden, I., *Nat. Geogr. Mag.*, 1944, **86**, 575–608.
78. Adams, M. R., in *Microbiology of Fermented Foods*, Ed. B. J. B. Wood, Elsevier Applied Science, London, 1985.
79. Freise, F. W., *Ind. Engng Chem. ind. Edn*, 1931, **23**, 1108–9.
80. Freise, F. W., *Dt. Essigind.*, 1931, **35**, 169–71.
81. Pilnik, W. and Voragen, A. G. J., in *The Biochemistry of Fruits and Their Products*, Vol. 1, Academic Press, London, 1970.
82. Francis, B. J. and Bell, J. M. K., *Trop. Sci.*, 1975, **17** (1), 25–44.
83. Elias, L. G., in *Coffee Pulp: Composition, Technology and Utilisation*, Eds J. E. Braham and R. Bressani, IDRC, Ottawa, 1979, 11–16.
84. Lane, A. G., *Fd Technol. Aust.*, 1979, **31**, 201–7.
85. Hobson, P. N., in *Processes for Chemicals from Some Renewable Raw Materials*, Institution of Chemical Engineers, London, 1979.
86. van Brakel, J., *Trop. Sci.*, 1980, **22** (2), 105–48.
87. National Academy of Sciences, Methane Generation from Human, Animal and Agricultural Wastes, NAS, Washington, DC, 1977.
88. Meynell, P. J., *Methane: Planning a Digester*, Prism Press, Dorchester, 1976.
89. Fry, L. J., *Practical Building of Methane Power Plants for Rural Energy Independence*, D. A. Knox, Andover, Hants, 1974.
90. Kirsch, E. and Sykes, R. M., *Prog. ind. Microbiol.*, 1971, **9**, 155–237.
91. Morgan, H., in *Food Industry Wastes: Disposal and Recovery*, Eds A. Herzka and R. G. Booth, Applied Science Publishers, London, 1981, 92–108.
92. Knol, W., Van der Most, M. M. and de Waart, J., *J. Sci. Fd Agric.*, 1978, **29**, 822–30.
93. Anon, *Coffee Trade News*, 1958, **9** (5), 20–1.
94. Anon, *Kenya Coffee*, 1967, **32**, 113–14.

95. Calle, H. V., *Indian Coffee*, 1957, **21**, 208–9.
96. Tiekjen, C., *Landb. Forsch-Volkenrode*, 1958 (1), 16–18.
97. Calzada, J. F., de Leon, O. R., de Arriola, M. C., de Miches, F., Rolz, C., de Leon, R. and Menchu, J. F., *Biotechnol. Letters*, 1981, **3** (12), 713–16.
98. Monteverde, F. and Olguin, E. J., in *Anaerobic Digestion and Carbohydrate Hydrolysis of Waste*, Eds G. L. Serriero, M. P. Feranti and H. E. Naveau, Elsevier Applied Science, London, 1984, 359–68.
99. Calle, H. V., *Bol. Tec. No. 6 Cenicafe, Colombia*, 1974, 11 pp.
100. Calzada, J. F., de Porres, E., Yurrita, A., de Arriola, M. C., de Micheo, F., Rolz, C. and Menchu, J. F., *Agric. Wastes*, 1984, **9**, 217–30.
101. Braham, J. E., in *Coffee Pulp: Composition, Technology and Utilisation*, Eds J. E. Braham and R. Bressani, IDRC, Ottawa, 1979, 51–4.
102. Cabezas, M. T., Flores, A. and Egana, J. I., in *Coffee Pulp: Composition, Technology and Utilisation*, Eds J. E. Braham and R. Bressani, IDRC, Ottawa, 1979, 25–38.
103. Bressani, R., in *Coffee Pulp: Composition, Technology and Utilisation*, Eds J. E. Braham and R. Bressani, IDRC, Ottawa, 1979, 83–8.
104. Jarquin, R., in *Coffee Pulp: Composition, Technology and Utilisation*, Eds J. E. Braham and R. Bressani, IDRC, Ottawa, 1979, 39–49.
105. Gomez Brenes, R. A., in *Coffee Pulp, Composition, Technology and Utilisation*, Eds J. E. Braham and R. Bressani, IDRC, Ottawa, 1979, 71–81.
106. Molina, M. R., de la Fuente, G., Batten, M. A. and Bressani, R., *J. Agric. Fd Chem.*, 1974, **22**, 1055–9.
107. Molina, M. R., in *Coffee Pulp: Composition, Technology and Utilisation*, Eds J. E. Braham and R. Bressani, IDRC, Ottawa, 1979, 63–9.
108. Barnett, A. J. G., *Silage Fermentation*, Butterworths, London, 1954.
109. Woolford, M. K., in *Microbiology of Fermented Foods*, Ed. B. J. B. Wood, Elsevier Applied Science, London, 1985.
110. Murillo, B., in *Coffee Pulp: Composition, Technology and Utilisation*, Eds J. E. Braham and R. Bressani, IDRC, Ottawa, 1979, 55–61.
111. Sivetz, M. and Desrosier, N. W., *Coffee Technology*, AVI, Westport, Conn., 1979, 519.
112. Blair, R., *Feedstuffs*, 30 June, 1975, 32–5.
113. Ravindramath, R., *J. Sci. Fd Agric.*, 1972, **23**, 307–10.
114. Anon, *El cafe de el Salvador*, 1958, No. 324–5, Nov.–Dec., 3.
115. Lane, A. G., *Biomass*, 1983, **3**, 247–68.
116. Sivetz, M., *Tea Coffee Trade J.*, 1977, August, 20–2.

Appendix

1. UNITS

1.1. SI Base Units

Quantity	Unit name	Unit symbol	Dimensions
Length	metre	m	[L]
Mass	kilogram	kg	[M]
Time	second	s	[T]
Thermodynamic temperature	kelvin	K	[θ]
Amount of substance	mole	mol	[N]
Electric current	ampere	A	[I]
Luminous intensity	candela	cd	[I_v]

1.2. Some SI Derived Units Used in Engineering

1.2.1. Units with Special Names

Quantity	Name	Unit symbol	Symbol expressed in base units
Frequency	hertz	Hz	s^{-1}
Energy, work, quantity of heat	joule	J	$kg\ m^2\ s^{-2}$
Force	newton	N	$kg\ m\ s^{-2}$
Pressure	pascal (newtons per square metre)	Pa	$kg\ m^{-1}\ s^{-2}$
Power	watt (joules per second)	W	$kg\ m^2\ s^{-3}$

1.2.2. Examples without Special Names

Physical quantity	SI unit	Unit symbol
Density	kilograms per cubic metre	kg m^{-3}
Heat capacity	joules per kilogram per kelvin	J kg^{-1} K^{-1}
Heat transfer coefficient	watts per square metre kelvin	W m^{-2} K^{-1}
Thermal conductivity	watts per metre kelvin	W m^{-1} K^{-1}
Velocity	metres per second	m s^{-1}
Viscosity		
(dynamic)	pascal second	Pa s
(kinematic)	square metres per second	m^2 s^{-1}

NB These unit symbols may alternatively be expressed using a solidus, e.g. W m^{-1} K^{-1} = W/m K.

1.3. Some Prefixes for SI Units

Multiplication factor	Prefix	Symbol
10^{12}	tera	T
10^{9}	giga	G
10^{6}	mega	M
10^{3}	kilo	k
10^{2}	hecto	h
10	deka	da
10^{-1}	deci	d
10^{-2}	centi	c
10^{-3}	milli	m
10^{-6}	micro	μ
10^{-9}	nano	n

NB The use of prefixes hecto-, deka-, deci-, and centi- is not recommended except for SI unit multiples for area and volume. The litre is an acceptable derived SI unit (equalling 1×10^{-3} cubic metres, or 1 cubic decimetre, or 1 dm^3), to which the above prefixes are commonly applied; thus, the millilitre (1/1000 or 10^{-3} litre), the centilitre (1/100 or 10^{-2} litre) and the decilitre (1/10 or 10^{-1} litre). The millilitre is equivalent in older use to the cubic centimetre (cm^3 or cc).

In weight and mass measurement, these prefixes are used in relation to the gram, e.g. the microgram (μg) is 10^{-6} g.

The common use of the millimetre, etc. in linear measurement is entirely consistent with basic SI units. The micron (often cited as μ, but more correctly as μm) is 1×10^{-6} m. The ångström is equal to 10^{-10} m; it is preferable to replace this unit with the nanometre (i.e. $1 \text{ Å} = 0.1$ nm).

1.4. Some Conversions of SI and Non-SI Units

Type of unit	To convert from	to	Multiply by
Linear	inch (in)	metre (m)	0.025 4
	foot (ft)	metre (m)	0.304 80
Area	square foot (ft²)	square metre (m²)	0.092 90
Volume	cubic foot (ft³)	cubic metre (m³)	0.028 32
	litre	cubic metre	10^{-3}
	gallon (US)	gallon (Imperial or British)	0.833
	gallon (British)	cubic metre	4.546×10^{-3}
Mass	pound (lb)	kilogram (kg)	0.453 6
	ounce (oz)	kilogram	28.35×10^{-3}
Density	pound per cubic foot	kilograms per cubic metre (or grams per litre)	16.02
		grams per cubic centimetre (or per millilitre)	0.016 02
Flow rate	cubic foot per minute (cfm)	cubic metre per second	4.72×10^{-4}
	gallon (British or Imperial) per hour (igph)	cubic metre per second	1.263×10^{-6}
Heat and power	Btu (BThU)	calories	252
	calories (cal)	joule (J)	
	thermochemical		4.184
	international table		4.187
	watt (W)	joule per second	equivalent
	horsepower (hp) (550 ft lbf per second)	watt (W)	7.457×10^{2}
	Btu per pound	{ calorie per kilogram	555
		{ kilocalorie per kilogram	0.555
Heat transfer coefficient	Btu per square foot per hour per °F	calorie per square metre per second per °C	1.356
	calorie per square metre per second per °C	watt per square metre per kelvin	4.187
Pressure	pound-force per square inch (psig, gauge; psig, absolute pressure)	{ kilogram per square centimetre	0.070 31
		{ pascal (Pa)	6895
		{ kilopascal (kPa)	6.895
		{ megapascal (MPa)	0.006 895
	inHg (32°F)	pascal	3.386×10^{3} ⎫
	mmHg (0°C)	pascal	1.333×10^{2} ⎬ *
	mmHg absolute	torr	equivalent ⎭
	bar	pascal	1.0×10^{5}
		megapascal	0.1
	atmosphere (=760 mmHg or ≈30 inHg ≈ 14.7 psi abs. = 0.00 psig)	pascal	1.013×10^{5}
		megapascal	0.101 3

* Also with gravity value of $9.806\,65\, \text{m}^2\,\text{s}^{-1}$.

Type of unit	To convert from	to	Multiply by
Thermal conductivity	Btu per hour per square foot per °F/in	calorie per second per square centimetre per °C/cm	3.447×10^{-4}
	Btu per hour per square foot per °F/ft		4.13×10^{-3}
	calorie per second per square centimetre per °C/cm (= calorie per second per centimetre per °C)	watt per square centimetre per kelvin/cm (= watt per centimetre per kelvin)	4.187
		watt per metre per kelvin	4.187×10^{2}
Viscosity (dynamic)	gram per centimetre per second (or dyne second per square centimetre)	poise (P)	equivalent
	poise	centipoise (cP)	10^{2}
	centipoise	pound (mass) per foot per hour	2.42
	gram per centimetre per second	pascal second	0.1
	centipoise	millipascal second (m Pa s)	equivalent
Viscosity (kinematic)	square centimetre per second	stokes	equivalent
	stokes	square metre per second	1.0×10^{-4}
Diffusivity (diffusion coefficient)	square centimetre per second	square metre per second	1×10^{-4}
Force	dyne	newton (N)	1.0×10^{-5}
	pound-force	newton (N)	4.448
Mass transfer flux	gram per square centimetre per second	gram per square metre per second	1×10^{4}

References

Kirk-Othmer, *Concise Encyclopaedia of Chemical Technology*, John Wiley, New York, 1985.

Stevens, C. B., *J. Soc. Dyers Colour.*, October 1978, 448–55.

1.5. Dimensionless Units Used

Unit (Number)		Symbol
Reynolds	$\dfrac{\text{(Diameter)(Velocity)}}{\text{(Kinematic viscosity)}}$	Re
Schmidt	$\dfrac{\text{(Diffusion coefficient)}}{\text{(Kinematic viscosity)}}$	Sc
Sherwood	$\dfrac{\text{(Mass transfer coefficient)(Diameter)}}{\text{(Diffusion coefficient)}}$	Sh

APPENDIX

Unit (Number)		Symbol
Prandtl	$\dfrac{\text{(Heat capacity)(Dynamic viscosity)}}{\text{(Thermal conductivity)}}$	Pr
Nusselt	$\dfrac{\text{(Heat transfer coefficient)(Diameter)}}{\text{(Thermal conductivity)}}$	Nu
Fourier	$\dfrac{\text{(Diffusion coefficient)(Time)}}{\text{(Diameter)}^2}$	Fo
Biot	$\dfrac{\text{(Mass transfer coefficient)}}{\text{(Diffusion coefficient)}}$	Bi

NB whereas, in the above units, diameter is twice the radius in the case of spheres, an equivalent diameter is used for other shapes.

Also, for a given temperature:

Physical quantity

Water activity
$\dfrac{\text{(Concentration of water vapour in air in equilibrium with given solution)}}{\text{(Concentration of water vapour in air in equilibrium with pure water)}}$ A_w
= Fractional relative humidity of air (RH)

Relative volatility (of component j) at infinite dilution in water (w) or other
$\dfrac{\text{(Activity coefficient)(Vapour pressure of pure component)}}{\text{(Vapour pressure of pure water)}}$ $\alpha^{\infty}_{j,w}$

Partition coefficient (air–water) of volatile component, j
$\dfrac{\text{(Concentration of component } j \text{ in unit volume of air)}}{\text{(Concentration of component } j \text{ in unit volume of water)}}$ k_j

Activity coefficient (of component j) at infinite dilution in water $\gamma^{\infty}_{j,w}$
Calculable from:

(i) $\dfrac{\text{(Partition coefficient)}}{(0.97 \times 10^{-6})\text{(Vapour pressure of pure component)}}$

(ii) $\dfrac{1}{\text{Mole fraction of component } j \text{ at saturation solubility}}$

(iii) Pierrotti correlations

and:

Mole-fraction of component, 1 in mixture with other components, $2 \to n$

$$\dfrac{\dfrac{\text{Weight of component, 1}}{\text{Molecular weight of component, 1}}}{\dfrac{\text{Weight of component, 1}}{\text{Molecular weight of component, 1}} + \cdots + \dfrac{\text{Weight of component, } n}{\text{Molecular weight of component, } n}}$$

2. SYMBOLS FOR PHYSICAL QUANTITIES IN EQUATIONS

For the specific use of symbols, see the context of particular equations; but, in general:

Physical quantity	*Symbol*
Pressure	P or p, with subscripts for partial pressure of particular components
Density	ρ, with subscripts of particular components
Viscosity	μ (dynamic); ν (kinematic)
Temperature	t, T or θ
Diffusion coefficient	D, with various subscripts referring to components, e.g. j, volatile compound; w, water, etc.
Heat transfer coefficient	h, with subscript referring to film phase in question; h is also used to signify porosity or interstitial void fraction. U is an overall heat transfer coefficient
Thermal conductivity	k, with subscript referring to a film phase (k_f) or material in question
Mass transfer coefficient	k (or k', for unidirectional transfer), with subscripts referring to film in question (e.g. g, w, l; or c (continuous), d (disperse phase)) or as an overall coefficient
Mass transfer flux	N or F, with subscripts and superscripts
Concentration	(or amount of solute in a given amount of a phase), C or ω, with varying use of particular superscripts and subscripts to indicate continuous or discontinuous phase, equilibrium conditions, dimensional/dimensionless quantities, initial and final conditions
Latent heat	ΔH_s (of fusion); λ (of evaporation)

3. ABBREVIATIONS

AFNOR	Association Française de Normalisation Tour Europe, Cedex 7, 92080 Paris La Defense, France
aq	Aqueous solution
ASIC	Association Scientifique Internationale du Café (all references to ASIC Colloquia use the date of the meeting, *not* the date of publication of the proceedings). Address: 42 rue Scheffer, 75016 Paris, France
as is	Composition based upon total weight of sample (i.e. no correction for water content)

BET equation Brunauer–Emmett–Teller equation (*J. Amer. Chem. Soc.*, 1938, **60**, 309) for estimating monolayer moisture content
BSI British Standards Institution, 2 Park Street, London, W1A 2BS
db Dry basis (i.e. corrected for water content)
EEC European Economic Community
ICO International Coffee Organisation, 22 Berners Street, London WC1P 4DD
ISO International Organisation for Standards, Case postale 56, CH-1211, Geneva 20, Switzerland
NTP At normal temperature (15°C) and pressure (760 mmHg absolute or 1·013 bar)
OAMCAF Organisation Africaine Mercantile de Café, comprises French-speaking countries of Benin, Cameroon, Central African Republic, Congo, Gabon, Ivory Coast, Madagascar and Togo within the ICO
STP At standard temperature (0°C) and pressure (760 mmHg absolute or 1·013 bar)
% w/w Weight of component as a percentage of total weight of sample

4. FLAVOUR TERMINOLOGY

The word 'taste' is often used interchangeably, but incorrectly from a technical viewpoint, with the word 'flavour'. A similar lack of distinction arises in the French language (with the words *le goût* or *la saveur*) and in German (with the single word *der Geschmack*). Flavour in the technical sense is the combined organoleptic sensation from basic taste (sour, sweet, bitter and salt), arising from the effect of non-volatile substances present on the tongue alone, together with astringency in the mouth; and from odour (comprising a variety of different 'notes') arising from the effect of volatile (to a greater or lesser degree) substances on the olfactory organs. In French, we have the words *l'odeur* or *le parfum*; and in German *der Geruch*.

With roasted coffee products, the particular word 'aroma' is used instead of 'odour', on account of the importance of its aromatic constituents based on ring structures (benzenoid, furanoid, etc.). Similarly, in French we use *l'arôme* or *les caractères aromatiques* (in contrast to *les caractères savoureuses*), and in German *das Aroma* (in contrast to *der Geschmack*).

The particular aroma involved may differ when assessed as either headspace aroma from dry roasted coffee or headspace aroma from a brewed cup of coffee, both directly sensed from the front of the nose, or as the component of flavour sensed primarily from within the mouth to the olfactory organs.

The vaporisation of the volatile compounds, and therefore greater capability for detection, is assisted by the procedure of 'slurping' during expert tasting. Coffee aroma comprises a very wide range of differently identifiable aromatic and non-aromatic notes, with different names (though without general agreement) evoked by association with other odours and flavours (e.g. 'floral', 'hay-like').

The term 'sensory [French *sensoriel*; German *sensorisch*] analysis' refers to the total evaluation by the human senses of a food product, such as a cup of coffee. This will include visual and organoleptic impressions, the latter including those of flavour and mouth feel.

Other adjectival terms in English, such as 'sapid', 'savoury' and 'gustatory', of uncertain technical meaning, are occasionally used.

5. PROCESS ENGINEERING TERMINOLOGY

5.1. Food Engineering and Unit Operations

Following the recognition of chemical engineering as a separate professional area of expertise (e.g. in the establishment of the American Institute of Chemical Engineers, and that of the Institution of Chemical Engineers (UK) in 1923), and later as a separate scientific discipline (e.g. in the establishment of university departments, notably at the Massachussetts Institute of Technology, USA), the concept of unit operations was developed by Arthur D. Little. This concept sought to show the unity of various physical operations, such as evaporation, mixing, etc., across a wide range of chemical and process industries, including food. A unit operation was defined as an operation in which a particular controlled physical change occurs in a material. The fundamentals of a given operation should then be studied in scientific, technical and economic terms, irrespective of the particular industry or material being handled. Where chemical changes which require control are also involved, the operation was less confidently defined as a unit process. Although many of the changes induced or taking place in foodstuffs (including coffee) are physical in nature, chemical changes are also evident; and furthermore also changes of a biochemical origin (e.g. action of enzymes and microorganisms). In recent years, the particularities of food processing have resurfaced as requiring separate emphasis, in the concept and study of food engineering as a derivative of chemical engineering (see Jowitt, R., *J. Food Engng*, 1982, **1**, 3–16; also Clarke, R. J., *Process Engineering in the Food Industries*, Heywood, London, 1957).

The concept of unit operations still serves a useful purpose, and is still conventionally used as a basis in many subject headings for texts devoted to process, chemical and food engineering, as in this present volume. There have been various attempts to classify unit operations in a logical manner, though none has been particularly successful, mainly on account of inevitable overlaps.

It is useful to note, however, that unit operations are essentially subdivisible into those involving (1) the separation of constituents (or molecular species) present in the same phase (whether gas, solid or liquid) or a mixture of different phases (whether present naturally or introduced, e.g. solvents); and those involving (2) the mixing of constituents in a given phase or mixture of phases. The separation of constituents comprises a very wide range of different techniques; thus we have simple physical operations such as filtration—the separation of particulate solids from liquids (or gases). Many other physical operations of separation, however, involve a change of phase for one or more constituents, such as evaporation, in which one or more constituents are vaporised from a liquid phase to cause differential separation (similarly, distillation), which is followed by the further operation of condensation, in which the separated constituents in the vapour phase are brought back to a separate liquid phase. Separations may also be brought about by use of membranes (e.g. reverse osmosis). This kind of analysis has led further to a concept of selectivity; that is, according to the particular unit operation being considered, there is a particular selectivity value for any given constituent over another, which will differ according to the physical conditions in the operation. As, in foodstuffs, the removal of water to cause concentration or dehydration is an important objective of processing, the selectivity of a given component is usefully related to that of the water molecules.

As a scientific discipline, chemical (and food) engineering has, however, lessened the significance of unit operations; but has developed the fundamental importance of phase equilibria (already inherent in 'static' physical chemistry, in particular thermodynamics) and of diffusion (mass transfer) of constituents through phases (inherent in 'dynamic' physical chemistry). Diffusion, applicable to both mixing and separation, becomes a primary determinant in the assessment of rates of change or occurrence, which is an essential factor in the design and operation of food and chemical processing plant. There are two other primary determinants with diffusion requiring fundamental study, those of momentum and heat transfer (inherent in physics).

Certain processes in the food industry are essentially chemical, such as roasting (notably in coffee), and many other essentially physical operations in the food industry are likely to have an important element of chemical and biochemical change, e.g. in drying and evaporation, and it is this element which has led to the particular desire for recognition of food engineering as a necessary derivative of chemical engineering.

Many operations in the food industry (and the coffee industry) are also, however, essentially mechanical (i.e. requiring mechanical forces alone); for example, subdivision without necessarily intending separation of constituents, as in grinding solids (though there may well be concomitant separation) or spraying liquids; other operations seek to separate different layers (as in 'peeling', 'shelling' and 'hulling').

This general discussion leads to brief definitions of other terms used, which may not be generally familiar.

Countercurrent. The contacting of two flowing streams of fluid (liquid or gas) containing constituent(s) which are to be separated, in opposing directions, since in this way the greatest average driving force (e.g. of concentration) is achieved in the entire equipment leading to maximum efficiency of separation. In *cocurrent* operation, the streams are contacted in parallel, and although in general less desirable, it may be necessary in practice where other factors are also important. These terms are also similarly used in relation to heat transfer.

Diffusion. The migration of a constituent(s) in a homogeneous material (either gas, liquid or solid) as a result of concentration variations from point to point, in such a way as to cause these concentrations to become uniform. Molecular diffusion is associated with the thermal agitation of individual molecules and their tendency to escape from a given area; eddy diffusion is a macroscopic movement within fluids, induced by mechanical or overall heat effects, assisting approach to uniform concentration.

Enthalpy. The heat content of a substance measured above a particular datum temperature (e.g. 0 or 40°C). In a solid substance, this heat content is the heat required to bring the unit weight of the substance from the datum temperature to the temperature at which the enthalpy is assessed (ambient or higher).

Equilibrium. A term particularly used in diffusional operations to express a condition in which the concentration of a particular constituent in one phase is constant with respect to its concentration in a second phase with which it is in intimate contact.

Extraction. Strictly speaking, solvent extraction, since a solvent (liquid) is used to extract a particular constituent(s) from a material (liquid or solid) in which it is preferentially soluble. Particularly in large-scale aqueous systems, the terms 'leaching' or 'lixiviation' are often used, whereas the terms 'infusion' and 'decoction' are used in small-scale extraction (e.g. 'coffee brewing'). Percolation is a particular technique of extraction in which the solvent (e.g. water) is allowed to pass through a fixed bed of particles of solid material from which the required constituent(s) are to be extracted.

Expression. A unit operation requiring only mechanical pressure to separate a liquid (containing required constituents) from an intimate solid–liquid mixture, often achieving a similar result to that of solvent extraction.

Material, heat and economic balances. Material and heat balances are the direct outcome of the laws of conservation of mass and energy, and are essential to the proper management of unit operations and processes in quantifying all material and heat inputs and outputs (including losses) at each stage. An economic balance, in its simplest form, is applied to unit operations and processes to provide a means of determining the point of lowest cost of operation, for example total (capital and running) costs of evaporation plotted against the number of effects (evaporation units) used.

Phase rule. $P + F = C + 2$; where P is the number of phases (gas, liquid or solid) present in a heterogeneous system; C is the number of components (as distinct molecular species); and F is the number of degrees of freedom in the system at equilibrium, that is, the number of elementary factors (i.e. temperature, pressure and concentration) that can be varied independently without altering the number of phases in the system.

Rate equations. A term particularly used in heat and mass transfer operations to express the rate of transfer in terms of the driving force (concentration difference in mass transfer), the area over which transfer is occurring, and a rate transfer coefficient.

Steady state. A term particularly used in heat and mass transfer, whether in countercurrent or cocurrent operation, where concentrations (temperatures in heat transfer) within each stream are constant with place (point) or time, which is most general in continuous operations (though not within the drying particles in spray-drying). Its converse, *non-steady* state, is inevitable in batch operations which are completed in finite time. The mathematics of non-steady-state operation is necessarily more complex than that of steady state.

Supercritical state. A fluid-phase condition of a substance, which is neither liquid nor gaseous, only achievable above a particular or critical pressure and temperature characteristic for the given substance (e.g. 75 bar and 31°C for carbon dioxide). It is a condition that confers enhanced solvent powers.

Transfer unit. A mathematical abstraction used in mass transfer calculations (continuous countercurrent) to express a localised unit (HTU, height or length) in which a particular level of separation (i.e. solute from one phase to another) is occurring (in certain cases, at equilibrium between a leaving and entering phase) in the equipment. A given number of such transfer units (NTU) are then necessary within the equipment to achieve a given total required separation. The product, HTU × NTU, gives the total height necessary.

6. LISTING OF BRITISH AND INTERNATIONAL STANDARDS RELATING TO COFFEE

Since 1963, the International Standards Organization, through a sub-committee (since 1975, TC34/SC15), has been developing a series of standards, referred to in the text, primarily of test methods relating to coffee and its products. The following is a listing of British Standards, together with the parent International Standard.

British Standard Number and Date	Title	Corresponding ISO Standard
Glossary		
BS 5456–1977	Glossary of terms relating to coffee and its products	ISO 3509–1984 Also amendment in DAM 3509–1983
Sampling of coffee and coffee products		
BS 6379		
Part 1 (1983)	Method of sampling green coffee in bags	ISO 4072–1982
Part 2 (1984)	Method of sampling instant coffee from cases with liners	ISO 6670–1983
Part 3 (1984)	Specification for coffee tryers	ISO 6666–1978

British Standard Number and Date	Title	Corresponding ISO Standard
Test methods for coffee and coffee products		
BS 5752		
Part 1 (1979)	Green coffee—determination of moisture content (basic reference method)	ISO 1446–1978
Part 2 (1984)	As above (practical method)	ISO 1447–1978
Part 3	Green, roasted and instant coffee—determination of caffeine content (modified Levine method)	ISO 4052–1983
Part 4 (1980)	Green coffee—olfactory, visual examination and determination of foreign matter and defects	ISO 4149–1980
Part 5 (1981)	Green coffee—size analysis by manual sieving	ISO 4150–1980
Part 6 (1984)	Instant coffee—determination of loss in mass at 70°C under reduced pressure	ISO 3726–1983
Part 7	Green coffee—determination of loss in mass at 105°C	ISO 6673–1984
Part 8 (1986)	Green coffee—determination of proportion of insect-damaged beans	ISO 6667–1985
Part 9 (1986)	Instant coffee—determination of insoluble matter	ISO 7534–1985
Part 10 (1986)	Instant coffee—size analysis	ISO 7532–1985
Others		
—	Green coffee in bags—guide to storage and transport	ISO/DIS 8455–1985
—	Woven wire cloth and perforated plate for control screens	ISO 565–1983

British Standard Number and Date	Title	Corresponding ISO Standard
BS 410 (1976)	Test sieves—technical requirements and testing	
Part 1	Test sieves of metal wire cloth	ISO 3310/1–1982
Part 2	Test sieves of metal perforated plate	ISO 3310/2–1982
BS 1796 (1976)	Method for test sieving	ISO 2591–1973

Index

Aagaard grader, 16
Abbreviations listed, 298–9
Acid content, brews, 232
Acidity, extraction yield relationship, 122
Activated sludge systems, 264–5
  deep-shaft process, 265
  tapered aeration system, 265
Africa hullers, 23, 24
Aged coffees, 51
Agglomeration, spray-dried powders, 148, 180–1
  process equipment for, 193
Air, physical properties of, 162
Air-conditioned stores, 52
Airline catering equipment, 225–6
Albro Fillers Company, 218
Aliphatic acids, roasted coffee content, 116
Alkaloids
  roasted coffee content, 116
  see also Caffeine
Amino acids
  roast coffee content, 116
  spent coffee grounds content, 285
Animal feed
  coffee pulp used as, 280–3
    maximum level used, 280–1
  husks used as, 259

Animal feed—*contd.*
  single-cell protein from coffee wastes, 273
  spent coffee grounds used as, 286
Aquapulper, 15
Araban, 120
Arabica coffees
  equilibrium relative humidity data, 50
  green bean shape, 39
  imports percentage data, 56
  processing of, 1, 3, 8
  typical compositions, 116
Aroma
  definition of, 299
  frosts, 140
  index (M/B ratio), 210
  sensory analysis of, 300
  stripping of, 68, 115
Ash content
  brews, 232, 233
  spent coffee grounds, 285
ASTM, screen mesh sizes, 100, 101
Atomiser, 149

Bag sizes, green coffee, 51–2
Biogas
  calorific value of, 278

Biogas—*contd.*
  equations for production, 278
  production from coffee pulp, 277–80
  sewage-generated, 278
Biological filter systems, 262–4
Biological oxygen demand (BOD)
  coffee processing wastewaters, 260
  river water, 261
Biot number, 127, 128, 131, 170, 297
Birs Technica spray-drying towers, 193
Black bean
  Brazilian/French/Spanish terminology for, 43
  equivalent count, 42–3, 44–5
  flavour effects of, 46
  origin of, 46
Blending methods, 57
Brazil
  coffee
    equilibrium relative humidity data, 50
    flavour characteristics, 38
  green beans
    colour classification system, 48
    defect grading system, 44, 45, 47
    harvesting times, 55
    processing method used, 1, 3
    size grades, 39, 42
Brew aroma, 140
Brewing control chart, 228, 229
Brews
  aromatic content of, 241, 243–4
  ash content of, 232, 233
  caffeine content of, 232
  carboxylic acid content of, 232
  chlorogenic acid content of, 232
  composition of, 231–3
  control chart for, 228, 229
  effect of additions, 228
  equipment for, 223–6, 234–6
  extraction yields for, 230–1, 238
  grind degree effects, 241, 242
  laboratory studies
    discussion of results for, 238–50
    experimental details for, 234–8
    objectives of, 234

Brews—*contd.*
  physicochemical characteristics of, 233
  polysaccharide content of, 232
  protein content of, 232
  raw materials effects on, 226–7
  sensory characteristics of, 230
  sensory evaluation of, 245–50
  solubilisation of individual constituents, 244–5
  soluble-solids content of, 239–40
  sugar content of, 232
  temperature effects on, 228–30
  trigonelline content of, 232
Brick packs, 214
British Standards, 304–6
  caffeine determination, 305
  defect grading, 305
  glossary of terms, 304
  instant coffee, 305
  sampling procedures, 304
  screen mesh sizes, 100, 306
  sieve analysis, 304, 305
  water content determination, 305
Bulk density
  free-fall measurement of, 104
  green beans, 4, 49, 85
  ground roast coffee, 104
  instant coffee, 147
  packed measurement of, 104
  roasted coffee beans, 85
  see also Density data
Burns Thermalo roasters, 90, 91, 92

Caffeine
  content in brews, 232
  instant/soluble coffee content, 59
  liquid coffee extract content, 59
  refining of, 69–70
    flow diagram for, 70
  roasted coffee content, 116
  uses of, 69
Caking, spray-dried powders, 201–2, 216
Cans
  ground roast coffee, 211–14
    advantages of, 212

Cans—*contd.*
  ground roast coffee—*contd.*
    carbon dioxide in, 213
    headspace in, 212–13
    inert atmosphere used, 213–14
    sizes used, 212
  instant coffee, 216
Caramelised products, 116, 117
Carbohydrates
  extraction yield relationship, 121–2
  roasted coffee content, 116
Carbon dioxide
  evolution of, in
    ground roast coffee, 99, 207–9
    roasted whole beans, 76–8, 202–5
    solutions to problem in packaging, 205
  retention of, in
    roasted beans, 77
  supercritical
    decaffeination by, 66–8
    volatiles extracted by, 136
Carboxy-5-hydroxytryptamides; *see* Coffee wax, C-5-HT fraction
Carboxylic acids
  content in various coffee products, 232
Catador pneumatic sorter, 27, 28
Catering equipment, 225–6
Cattle, coffee waste products fed to, 259, 280–1, 286
Ceka container system, 218
Centritherm evaporator, 137, 194
Chaff, 77
  grinding release of, 99
Charcoal, waste products converted to, 259, 277
Charring, roasted coffee, 79
Cherco (animal feed), 286
Chickens, fermented pulp fed to, 281
Chicory, calorimetric curve for, 79
Chlorogenic acids (CGA)
  content in brews, 232
  roast degree estimated by analysis, 87
  roasted coffee content, 116
Cleaning processes, 22, 52–3

Cocurrent operations
  meaning of term, 302
  spray-drying, 156, 165
Coffee Brewing Institute, brewing control chart, 228, 229
Coffee flakes, 98, 115, 195
Coffee oil
  content in various coffees, 116
  purification of, 139
  recovery of, 138–9
  reincorporation into instant coffee, 139–41, 216
  volatiles retention affected by, 172
Coffee wax, C-5-HT fraction, effect of dewaxing on, 53
Coffex AG patent, 66
Colburn equation, 128–9
Collapse temperature concept, freeze-dried coffee, 183–4
Colorimetric sorting
  electronic equipment, 29–31
  manual methods, 29
Colour
  changes during roasting, 84–5
  classification, green beans, 48
  coffee brews, 233
  measurement of, 86
Columbia
  coffee
    flavour characteristics, 38
  green beans
    harvesting times, 55
    size grade, 39
Compositional data
  brews, 231–3
  roast coffee, 116–17
Composting (of wet processing pulp), 269–71
  carbon/nitrogen ratio for, 270–1
  moisture content limits, 270
  particle size requirements, 270
  use of end-product, 271
  windrow system, 270
Cona brewing equipment, 225, 235
  see also Continuous water-circulation brewing equipment
Condensation products, 116, 117

Consumption, decaffeinated coffee, USA, 60, 61
Continuous countercurrent extractors, 110, 142–3
  extraction productivity for, 132, 133
  mass balance calculation for, 124
  mass transfer rate equations for, 130–1
  yield data, 125, 126
Continuous water-circulation brewing equipment, 225, 234–5
  extraction yield data, 238, 240
  sensory evaluation of brews, 249, 250
  soluble-solids content of brews, 240
  volatiles lost in, 241, 243
Costa Rica, green bean harvesting times, 55
Countercurrent operations
  extraction, 110, 142–3
  meaning of term, 302
Crop year, 49
  definition of, 54
  grade designation, 49
Cryogenic grinding, 99
Cup testing, procedure, 38
Curing operations
  cleaning, 22
  colorimetric sorting, 28–31
  density sorting, 27–8
  hulling, 22–6
  redrying, 22
  size grading, 26

Decaffeinated coffee
  carbon dioxide evolution from, 203, 208
  consumption in USA, 60, 61
Decaffeination
  green beans, 61–8
  invention of process, 59
  roasted coffee extracts, 68–9
  solvent processes, 61–4
    basic principles, 62
    criteria for choice of solvent, 61
    flow diagram, 63

Decaffeination—*contd.*
  solvent processes—*contd.*
    operating conditions for, 63, 64
    water content minimum, 62–3
  solvents used, 61–2
  supercritical carbon dioxide processes, 66–8
    advantages/disadvantages, 66–7
    flow diagram, 67
    process conditions, 68
  water process, 64–6
    caffeine recovery operations, 65–6
    flow diagram, 65
Decoction technique (of brewing), 224
Defect grading systems, 42–8
  extraneous matter ratings, 43, 45
  terminology, 43
  type numbering systems, 46, 47
  various countries, 43, 44–5
  weight basis, 46
Degassing
  ground roast coffee, 213–14, 215
  process equipment for, 217–18
  roast whole beans, 207
Degree of roast; *see* Roast degree
Density data
  green beans, 4, 49, 85
  see also Bulk density
Density sorting
  gravimetric sorters, 28
  pneumatic sorters, 27–8, 29
Descascador huller, 26
Destoning processes, 22, 52–3
Dewaxing processes, 53
Diffusion, meaning of term, 302
Diffusivity, liquid coffee extracts, 165–7
Dimensionless numbers, 296–7
Disc pulpers, 10–11, 12, 13
Drip-pot method (of brewing), 223–4
Drum driers, 147
Drum pulpers, 12–13
Dry processing, 3–8, 9
  artificial drying methods, 4–8
    furnace construction for, 6–7
    rotary drum driers, 7–8
    static driers, 7
    vertical driers, 8, 9

Dry processing—*contd.*
  natural drying method, 3–4
  waste products from, 257–9
Drying processes
  freeze-drying, 181–9
  fuel for, 6, 259
  green coffee beans, 3–8
  liquid coffee extracts, 147–96
    process types, 147
    see also Freeze-drying; Spray-drying
  parchment coffee, 19–22
  pre-concentration methods, 148, 190–2
    evaporation methods, 190
    freeze-concentration method, 190–2
    reverse osmosis, 192
  process control of, 195–6
  pulp waste product from, 282
  process equipment for, 192–6
  spray-drying, 149–79
Duplex pneumatic sorter, 27, 29
Dust hazards, spray-driers, 196

El Salvador, wastewater treatment plant, 273
Electronic sorting equipment, 29–31
  thoroughness of sort calculated, 31
Emission control, roaster exhaust fumes, 87–9
Engelberg-type hullers, 23, 24
Ensiled coffee pulp, animal feed use of, 282–3
Enthalpy
  data for green coffee beans, 80
  meaning of term, 302
Entoleter mill, 106
Equilibrium, meaning of term, 302
Equilibrium relative humidity (ERH)
  values quoted for green coffee, 50
Espresso machines, 226
  household versions, 225, 236
  particle size of coffee grinds, 98
  see also Vaporisation-under-pressure brewing equipment

Ethanol
  production from coffee pulp, 274
  world production data, 274
  yield from carbohydrates, 274
Ethiopia
  green bean defect grading system, 44, 45, 47
Evaporation
  pre-concentration processes, 190
  process equipment for, 194
  volatile stripping processes, 136–7
Evaporators, 194
Excelso size grade, 39
Exports, timing of, 55
Expression
  coffee oil, 138–9
  meaning of term, 303
Extraction processes, 109–44
  compositional factors, 117–22
  countercurrent continuous screw extractors, 110, 142–3
  definition of, 222, 303
  draw-off factor defined, 124
  final state in, 223
  green yield, 118
  initial state in, 222–3
  mass and heat balances for, 122–4
  mass transfer rates, 124–32
    fixed-bed equations, 129–30
    intra-particle diffusivity equations, 127–8
    liquid film resistance equations, 127
    lumped resistance model equations, 128
    rate equations, 126–32
  meaning of term, 222, 303
  mechanisms
    soluble solids extraction, 114–15
    volatile compounds extraction, 115
  methods for, 110–14
  monosaccharide content data, 119–20
  percolation batteries, 111–13, 141–2
    pressurised system, 111, 114
    split draw-off concept for, 113
    temperature effects on, 111, 114

Extraction process—contd.
  process control measurements for, 143–4
  process equipment, 110–14, 141–3
  productivity factors for, 132–3
  slurry systems, 113–14
  transfer period of, 223
  unit operations treatment of, 114, 122–32
  water, effects on, 134
  yield data
    autoclave extraction, 119
    household extraction, 118
    industrial process, 119, 125, 126
    temperature effects on, 119–20
  see also Solid–liquid extraction processes
Extraneous matter
  black-bean-equivalency ratings for, 43, 45

Fair average quality (FAQ) designation, 38, 47
Fast-roasted coffee, 74, 78
Fatty acids
  spent coffee grounds content, 285, 286
Fermentation processes, 17–18
Fick's laws of diffusion
  carbon dioxide evolution, 204
  drying processes, 163, 164, 173
  extraction processes, 126
Film evaporators, 190, 194
Filter brewing equipment, 224, 235–6
  extraction yield data, 238, 240
  particle size of coffee grinds, 98
  sensory evaluation of brews, 249, 250
  soluble-solids content of brews, 239
  volatiles retained in, 243
Fines, spray-dried powder, 179–80
Fire hazards, spray-driers, 196
Fitzmill mill, 106
Flaked coffee, 98, 115, 195
Flats, 39
Flavour
  changes in coffee extracts, 122

Flavour—contd.
  characteristics, specification of, 38–9
  instant coffee, 141
  terminology for, 299–300
Fluidised-bed equipment
  driers, 147
  roasters, 96–7
    air-to-coffee weight ratio for, 84, 96
Food engineering, 300–1
Fourier number, 127–8, 131, 170, 174, 175, 297
France
  arabica/robusta imports proportion, 56
  green bean defect grading system, 44, 45
Free amino acids (FAA)
  green coffee, 116
  roast coffee, 116
  see also Amino acids
Freeze-concentration, 190–2
  ice separation methods for, 191
  process equipment for, 194
Freeze-dried instant/soluble coffee
  collapse temperature concept, 183
Freeze-drying
  drying rate equations, 186
    factors affecting, 187
  first done commercially for coffee, 148
  heat requirement for, 184
  heat-transfer-controlled processes, 185, 186
  ice-front temperature profiles, 185–6
  mass-transfer-controlled processes, 185, 186
  methods for, 181–2
  process factors in, 181–9
  temperature used, 183
  volatiles retention in, 188–9
    factors affecting, 189
  water-removal mechanism for, 182–7
Fuel
  husks used as, 6, 259
  parchment used as, 277
  spent coffee grounds used as, 286–7

Galactan, 120
Gardner (automatic colour difference) meter, 86
General Foods Corporation decaffeination processes, 60, 64
Germany
  arabica/robusta imports proportion, 56
  emission laws, 88, 89
  health coffees, 53
Glossary of terms
  flavour, 299–300
  process engineering, 302–4
  Standards for, 304
Glucan, 120
Gothot Rapido-Nova batch roaster, 93–5
  emission volatiles from, 88
  heat requirements for, 82–3
Graders
  Aagaard densimetric grader, 16
  rotatory drum, 26
Grading
  bean size/shape, 39–42
  bulk density, 49
  colour, 48
  crop year, 49
  defects, 42–8
  flavour characteristics, 38–9
  roasting characteristics, 48–9
Granulizer, 105, 106
Gravimetric sorters, 28
Green coffee beans
  calorimetric curve for, 79
  exothermic reactions in, 79
  floating/non-floating distribution, 4
  processing methods, 1–22
    curing operations, 1, 2, 22–31
    dry processing methods, 1, 2, 3–8
    flow sheet showing, 2
    grading operations, 26
    handling methods, 32–3
    sorting operations, 27–31
    storage methods, 31–2
    wet processing methods, 1, 2, 8, 10–22
    see also Dry...; Wet processing
  size distribution, 5

Green coffee beans—*contd.*
  specific heat data, 80
Grind degree, 98
  brews affected by, 227, 241, 243
  methods of assessing, 99–104
  screen analysis for, 99–104
Grinder gas, 99, 140
Grinding
  cryogenic processes, 99
  equipment for, 104–6
  mechanism of, 97–9
  particle size analysis, 101, 102
    coefficient of variation of, 102
  particle size requirements, 98–9
  process factors in, 97–104
Ground roast coffee
  bulk density of, 104
  carbon dioxide evolution from, 99, 207–9
  packing of, 207–15
  preparation of, 97–9, 104–6
  size analysis of, 99–104
Grounds; *see* Spent coffee grounds
Guardiola drier, 7, 8
Gump grinder, 105, 106

HAG–GF decaffeination process, 66
Handling methods, green beans, 32–3
Hard coffee, meanings of term, 3, 29
Hard packs, 214
Harvesting times, 55
Headspace aroma, 139, 299–300
  changes in packed coffee, 211
Health coffees, 53–4
Heat balance calculations
  extraction process, 124
  meaning of term, 303
  roasting process, 81–2
  spray-drying, 157–8
Heat requirements, roasting processes, 80–3
Heat transfer rate calculations
  roasting processes, 83–4
  spray-drying process, 161–5
High-temperature drying, flavour effects of, 5

INDEX

Home brewing
 bibliographic review of, 222–33
 equipment for, 224–5, 234–6
 extraction yield data, 118
 laboratory studies on, 234–50
 manual techniques for, 223–4
 raw materials for, 226–7
Household equipment, coffee brewing, 224–5, 234–6
Hulling equipment
 Africa hullers, 23, 24
 Descascador huller, 26
 Engelberg-type hullers, 23, 24
 Okrassa huller–polisher, 25
 Smout huller–polisher, 23, 25
Humic acids, roast coffee content, 116
Hunterlab meter, 86
Husks
 animal feed use for, 259
 composition of, 258
 fuel use for, 259

Ice, water vapour
 pressure–temperature curve, 183
Immature beans
 black-bean equivalency ratings for, 44
 Brazilian/French/Spanish terminology for, 43
Imports, arabica/robusta proportions for various countries, 56
Indonesia
 green bean defect grading system, 44, 45, 47
Infusion method (of brewing), 224
Insect damage, green beans in storage, 32
Instant coffee
 bulk density of, 147
 caffeine content of, 59
 caking of, 201–2, 216
 packing of, 215–16
 physical form of, 147
 sales proportion of, 57
 selection factors for, 56–7
 volatiles reincorporated into, 139–41, 216

Instant coffee—*contd.*
 waste products from manufacture of, 283–7
  composition of, 285
  recovery process for, 283, 284
  use as animal feed, 286
  use as fuel, 286–7
 see also Spent coffee grounds
Instantising (of milk powders), 148, 180
International Standards, 304–6
 brewing procedure for cup testing, 38
 colour classification of green beans, 48, 305
 defect grading procedures, 46, 48, 305
 glossary of terms, 304
 instant coffee, 305
 sampling procedures, 37, 304
 screen analysis procedure, 40, 305
 screen mesh sizes, 100, 305–6
 water content determination, 305
Intra-particle diffusivity
 extraction process, 127–8
 spray-drying process, 163
Irrigation, wastewater used for, 268
Isotherms
 water content of green coffee beans, 50
 see also Sorption isotherms; Water content
Italy, arabica/robusta imports proportion, 56
Ivory Coast
 fermentation equipment, 17
 green beans
  defect grading system, 46, 47
  harvesting times, 55
  size grades, 39, 41
  storage methods, 32

Jigging operations, green bean grading, 16
Jute bags, 51, 52

Kaffee Hag, 59
Kahawa coal, 277

Kellogg Company, 59, 60
Kenya
  coffee
    equilibrium relative humidity data, 50
    flavour characteristics, 38
    green beans
      colour classification system, 48
      harvesting times, 55
      size grades, 39
      sun drying methods, 21–2
    parchment charcoal product, 277
    river pollution in, 261
    wastewater disposal in, 267
Kisher (*ghishr*), 275
Kivu pump, 19

Land disposal (of wastewaters), 267–8
Lendrich steaming process, 54
LePage bean-cutting rolls, 105, 106
Light reflectance meters, 86
Lignin
  roast coffee content, 116
  spent coffee grounds content, 285
Liquid coffee extracts
  acidity of, 122
  diffusion coefficients for, 166, 169, 170
  equilibrium diagrams for, 131
  monosaccharide content of, 119–21
  refractive index measurement of, 143
  soluble solids content of, 110, 148
  specific gravity measurement of, 143
  stripping of volatiles from, 136–7
  viscosity–solids content relationship, 190
  water activities in, 166
Liquoring characteristics, specification of, 38–9
London Terminal Market Robusta (LTMR)
  defect grading system, 44, 45, 47
Lurgi Aerotherm roaster, 96

Maltodextrin solutions
  critical moisture contents of volatile compounds in, 169
  diffusion coefficients for, 166, 169, 170
  water activities in, 166
Maltose solutions
  diffusion coefficients for, 166, 169, 170
  water activities in, 166
Mannan, 120
Margaropipes, 39
  selection of, 56
Marketed brands, blending for, 56
Marketed grades, specification systems for, 36–8
Mass balance calculations
  extraction process, 122–4
  meaning of term, 303
  roasting process, 81–2
  spray-drying, 158–9
  wet processing, 261
Mass transfer rate calculations
  spray-drying process, 161–5
Melanoidins, roast coffee content, 116
Methanogenic bacteria, pH sensitivity of, 279
Methylxanthines
  organoleptic properties of, 230
  transfer to brew of, 232
  see also Caffeine
Mexico, coffee
  harvesting times, 55
Mild coffee, processing method resulting in, 1
Minerals, roasted coffee content, 116
Mocha coffees, selection of, 56
Moisture content
  green coffee beans, 4, 21, 22, 31, 51
  roasted beans, 87
  sample for, 37
Monosaccharides
  coffee extract content, 120–1
  extraction temperature effect on, 120
  see also Sugars
Monsooned coffees, 51
Moreira drier, 8

Mould growth, minimum moisture content for, 31

New York Coffee and Sugar Exchange (NYCSE) defect grading system, 42–3, 44, 45, 48
Nicotinic acid
 roasted coffee content, 116
 roasting, effects on, 233
Niro Atomizer A/S equipment
 continuous countercurrent extractor, 142–3
  yield relationships for, 126, 132, 133
 spray drier, 192, 193
Non-return valves (for CO_2 release), 205, 207
Normalising, 99, 106
Nusselt number, 161, 165, 177, 297

Okrassa equipment
 drier, 7
 huller–polisher, 25
Organoleptic properties, coffee brew, 230

Packing
 ground roast coffee
  cans, 211–14
  carbon dioxide evolution, 207–9
  hard (laminated) packs, 214–15
  packing equipment used, 218
  packs used, 211–15
  soft packs, 215
  stability factors for, 209–11
 instant coffee, 215–16
  containers used, 216
  packing equipment used, 218
  problems encountered, 201–2
 roast whole bean coffee
  carbon dioxide evolution in, 202–5
  packs used, 206–7
  stability factors for, 206
 weight control systems, 218–19

Palletisation, bagged green coffee, 52
Parchment coffee
 disposal of waste, 277
 drying of, 19–22
Particle size measurement
 grind degree assessed by, 99–104
 laser beam instruments, 104
 screen analysis methods, 101–4
 see also Screen analysis
Peaberries, 39
 selection of, 56
Pectin
 production from coffee pulp, 275–7
 roasted coffee content, 116
 uses of, 276
 world production data, 276
Percolating filters, wastewater treatment by, 262–4
Percolation batteries, 111–13, 141–2
 extraction productivity for, 132–3
 mass balance for, 123
 split draw-off concept, 113
 temperature effects, 111, 114
Percolators, 224–5, 235, 236
 extraction yield data, 238, 240
 particle size of coffee grinds, 98
 sensory evaluation of brews, 249, 250
 soluble-solids content of brews, 240
 volatiles lost in, 241, 243
Phase diagrams
 liquid coffee extracts, 131, 182, 184
Phase rule, meaning of term, 303
Phenolic acids, organoleptic properties of, 230
Photovolt meter, 86
Physicochemical characteristics, coffee brews, 233
Pillow/pouch packs, 215
Plating oils, 139, 140
Pneumatic Scale Corporation, 218
Pneumatic sorters, 27–8, 29
Pod, black-bean-equivalency ratings for, 44
Polishers, 23–5
Polysaccharides
 content in brews, 232
 extraction yield relationship, 121–2
 roasted coffee content, 116

Portugal, green bean defect grading system, 44, 45
Prandtl number, 161, 297
Pre-concentration, 148, 190–2
　evaporation processes, 190
　freeze-concentration processes, 190–2
　equipment, 194
　reverse osmosis process, 192
Pressurised roasters, 97
Pre-treatments
　cleaning/destoning, 52–3
　dewaxing, 53
　steam treatment, 54
Probat roasters, 89–90, 91, 92, 95
Process engineering terminology, 300–4
Process equipment
　caffeine refining, 70
　decaffeination, 63, 65, 67
　drying, 192–6
　extraction, 141–4
　green bean processing, 4–30
　grinding, 104–6
　roasting, 74, 89–97
Proteins
　content in brews, 232
　roasted coffee content, 116
　spent coffee grounds content, 285, 286
Pulp
　animal feed uses, 280–3
　　detoxification effects, 281
　　maximum level used, 280–1
　biogas produced from, 277–80
　composting of, 269–71
　definition of, 10
　drying of, 282
　ensiling for animal feed use, 282–3
　ethanol produced from, 274
　microbial biomass produced from, 271–4
　pectin produced from, 275–7
　vinegar production from, 275
　wine produced from, 274–5
Pulping operations
　disc pulpers, 10–11, 12, 13
　drum pulpers, 12–13

Pulping operations—*contd.*
　pulper–repasser system, 14, 15
　Raoeng pulper, 14–15, 16
　roller pulper, 15

Quenching, 75
Quinic acid, roasted coffee content, 116

Ranz–Marshall correlations, 160, 165
Raoeng pulper, 14–15, 16
Rate equations
　meaning of term, 303
　roasting, 83–4
　spray-drying, 161–5
Reverse osmosis, 192
Reynolds number
　definition of, 296
　extraction process, 131
　spray-drying process, 160–1, 162, 165
Roast degree
　dry matter loss affected by, 75
　effects of, 74
　measurement of, 86–7
Roasted coffee beans
　aliphatic acids content, 116
　amino acids content, 116
　caffeine content, 116
　caramelised products content, 116, 117
　carbohydrates content, 116
　carbon dioxide evolution from, 76–8, 202–5
　chlorogenic acids content, 116
　home brewing use of, 226–7
　lignin content, 116
　mineral content, 116
　nicotinic acid content, 116
　oil content, 116
　packing of, 202–7
　pectins content, 116
　polysaccharides content, 116
　proteins content, 116
　quinic acid content, 116
　soluble solids data for, 78

Roasted coffee beans—*contd.*
  sugars content, 116
  trigonelline content, 116
  typical compositions, 116–17
  volatiles content, 116
Roasters, 74, 89–97
  air-to-coffee weight ratio for, 84, 96
  ancillary equipment for, 97
  batch roasters, 89–90, 91, 93–5
  continuous roasters, 90–2
  emission control for, 87–9
  fluidised-bed roasters, 74, 96–7
  gas-to-coffee weight ratios for, 83, 84
  horizontal drum roasters, 74, 89–92
    batch drum roasters, 89–90, 91
    continuous roasters, 90–2
  pressurised roasters, 97
  recirculating roasters, 90, 92, 94
    heat factors affected by, 82, 83
  rotating-bowl roasters, 74, 95
  vertical static drum with paddles, 74, 93–5
Roasting processes
  carbon dioxide, evolution from, 76–8, 202–5
  chemical changes during, 75–9
  density changes during, 85
  dry matter loss during, 75–6
  energy requirements for, 80–3
  gases from, 76
  heat factors in, 79–84
  internal structural changes during, 85–6
  mass and heat balance calculation, 81–2
  mechanical principles in, 74
  mechanisms of, 73–4
  physical changes in, 84–6
  pollution control of, 75
  soluble solids data for, 78
  temperature effects on, 78–9
Robusta coffees
  brittleness of green beans, 32–3
  equilibrium relative humidity data, 50
  green bean shape, 39
  imports percentage data, 56

Robusta coffees—*contd.*
  processing of, 1, 3
  typical compositions, 116
Roselius, Ludwig, 59, 60
Roasting characteristic, green bean grading, 48–9
Rotap sieving machine, 103
Rotary drum driers, 7–8
Rotating biological contactor (RBC), 263–4

Sänft coffees, 53
Sanka Coffee Corporation, 60
Schmidt number, 131, 160, 161, 296
Screen analysis
  air-jet deblinding device for, 103
  equipment suppliers for, 40
  examples
    Brazilian arabica, 42
    Ivory Coast robusta, 41
  grind degree assessed by, 99–104
  mesh sizes, 100–1
  numbering system, 40
  results for coffee grinds, 101–4
  Standards for, 40, 305
  test procedure, 40
Seasonal availability, 54
Seepage pits, 267
Selection factors, 54, 56–7
Selective diffusion concept, 167–71
  experimental trials of, 177–9
  rate equations for, 172–7
Sensory analysis
  brews, 237–8, 245–50
    canonical analysis of, 247–50
    graphical representation of, 245–7
  definition of, 300
Shape grading, 39
Shelf-life, packed coffee, 210
Sherwood number, 74, 131, 160, 165, 296
SI units
  base units, 293
  conversion factors for, 295–6
  derived units, 293–4
  prefixes for, 294–5
Silos, 52

INDEX

Silverskin
  chaff resulting from, 77
  disposal of, 277
Single-cell protein (SCP)
  consumer acceptability of, 273
  production from coffee pulp, 271–4
    economics of, 273
    yield of dry product, 272
Size grading
  equipment, 16, 26
  laboratory test screens, 40–1
  specifications, 39–42
Smout huller–polisher, 23, 25
Solid–liquid extraction processes, 222–3
  see also Extraction processes
Soluble coffee
  sales proportion of, 57
Soluble solids content, roasted coffees, 78
Soluble solids extraction, mechanism of, 114–15
Solvent decaffeination, 61–4
Sortex electronic sorter, 29–30
Sorting equipment
  Catador pneumatic sorter, 27, 28
  colorimetric sorters, 28–31
  density sorters, 16, 27–8, 29
  Duplex pneumatic sorter, 27, 29
  electronic sorters, 29–31
    thoroughness of sorting calculated, 31
  gravimetric sorters, 28
  pneumatic sorters, 27–8, 29
  Sortex electronic sorter, 29–30
Sour bean
  black-bean-equivalency ratings for, 44
  Brazilian/French/Spanish terminology for, 43
  flavour effects of, 46
  origin of, 46
Spain, arabica/robusta imports proportion, 56
Specific heat data, green coffee beans, 80
Specifications, 36–8
  characteristics of, 36

Spent coffee grounds
  calorific value of, 285, 287
  composition of, 285
  recovery process for, 283, 284
  use as animal feed, 286
  use as fuel, 286–7
Spray-drying
  agglomeration of powder, 148, 180–1
  centrifugal pressure nozzles used, 152–5
    flow number (FN) calculation, 153
  compositional changes during, 150–1
  diffusivity considerations, 165–7
  droplet formation in, 151–5
  dust hazards in, 196
  external variables in, 178
    pilot plant trial values, 179
  fines separation in, 179–80
  fire hazards in, 196
  methods for, 149–50
  physical properties of air used, 162
  pneumatic sprayers used, 155
  process equipment for, 192–3
  process factors in, 149–79
  rate factors, short-cut calculation method, 164–5
  retention of volatiles in, 167–79
    experimental trials, 177–9
    factors maximising retention, 176
    non-diffusional loss, 171–2
    rate factor equations, 172–7
    selective diffusion concept, 167–71
  spinning discs used, 151, 155
  spray–air contact in, 155–6
  spray cone angle for, 151, 154
  volatiles loss during, 151
  water-removal mechanisms for, 156–67
    mass and heat balances, 157–9
    rate factors, 159–67
Stabilisation (wastewater treatment) ponds
  anaerobic ponds, 266–7
  facultative ponds, 266, 267

Stability factors
  green beans, 50–1
  ground roast coffee, 209–11
  roasted whole beans, 206
Staling (of ground roast coffee), 206
  oxygen requirement for, 211
Steady-state operations, meaning of term, 303
Steeping procedure, 38
Stinker beans
  Brazilian/French/Spanish terminology for, 43
  formation of, 5, 46
Storage
  green beans, 31–2, 49–52
    conditions required, 49–50
    methods of storage, 32, 51–2
    stability criteria, 50–1
Sucrose
  roasted coffee content, 116
Sugars
  content in brews, 232
  roasted coffee content, 116
Sun drying methods
  green berries, 3–4
  parchment coffee, 21–2
  pectin product from pulp, 277
  pulp waste product, 282
Supercritical carbon dioxide
  decaffeination by, 66–8
  volatiles extracted by, 136
Supercritical state, meaning of term, 304
Supremo size grade, 39
Sweden, arabica/robusta imports proportion, 56
Switzerland, arabica/robusta imports proportion, 56
Symbols, equations, 298

Tanzania
  coffee, flavour characteristics, 38
  green bean defect grading system, 47
Tar deposits, roasted coffee, 79
Temperature effects, coffee brews, 228–30
Torres drier, 7

Transfer units, meaning of term, 304
Trickling filters, wastewater treatment by, 262–4
Trigonelline
  content in brews, 232
  roasted coffee content, 116
  roasting, effects on, 87, 233
Turkish-type coffee
  grind size for, 99
  method of preparation, 224
Tyler series screens, 101, 102

UK
  arabica/robusta imports proportion, 56
  defect grading system, 44, 45, 47
  standards; *see* British Standards
Unit operations
  concept first developed, 300
  extraction as, 114, 122–32
  types of, 301
Units, 293–4
  conversion factors for, 295–6
  prefixes for, 294–5
Urns, 225
USA
  arabica/robusta imports proportion, 56
  consumption per head, 60
  decaffeinated coffee consumption, 60, 61
  green bean defect grading system, 42–3, 44–5
Usual good quality (UGQ) designation, 38, 47

Vacuum driers, 147, 195
Vacuum method of brewing, 225, 234–5
Vacuum packs, 205, 207, 211–14
Vaporisation-under-pressure brewing equipment
  extraction yield data, 238, 240
  sensory evaluation of brews, 249, 250
  soluble-solids content of brews, 239
  volatiles in, 243

INDEX 321

Vending machines, 226
Vertical driers, 8, 9
Vertical washers, 19, 20
Vinegar, production from coffee pulp, 275
Volatile components
  condensation and collection of, 137–8
  diffusivity in liquid coffee extracts, 168–9
  extraction of, 115, 136
  loss during condensation, 138
  prestripping from ground roast coffee, 134–6
  reincorporation into instant coffee, 139–41, 216
  steam stripping from ground roast coffee, 134–5
  stripping from extracts, 136–7

Wastewater oxidation systems, 260–6
  activated sludge systems, 264–5
  biological filters, 262–3
  drawbacks of, 265–6
  rotating biological contactors (RBCs), 263–4
Water
  activity
    liquid coffee extracts, 165–7
    relationship with water content, 167

Water—*contd.*
  decaffeination systems, 64–6
  effects on brew, 227
  extraction affected by, 134
Weight control, packed coffee, 218–19
Wet processing, 8, 10–22
  disadvantages of, 259–60
  draining/pre-drying operations, 19
  drying of parchment coffee, 19, 21–2
  fermentation processes, 17–18
  grading operations, 16
  pulping operations, 10–16
  receiving operations, 10, 11
  separation/classification methods, 15–16
  washing operations, 18–19
    manual methods, 18–19
    mechanised washers, 19, 20
  waste products from, 258, 259–83
    land disposal of, 267–8
    oxidation systems, 260–6
    pollution data, 260, 262
    precipitation of, 268–9
    stabilisation ponds for, 266–7
  water consumption for, 19, 260
Wine, production from coffee pulp, 274–5
Wiped film evaporators, 190, 194
Wolverine Jet Zone roaster, 96, 97

Printed in Poland
by Amazon Fulfillment
Poland Sp. z o.o., Wrocław